Comedia

Series editor: David M

Componer

Components of Dress

Design, manufacturing, and image-making in the fashion industry

Edited by

Juliet Ash and Lee Wright

A Comedia book
published by Routledge
London and New York

A Comedia book
first published in 1988 by
Routledge
a division of
Routledge, Chapman
and Hall
11 New Fetter Lane, London EC4P 4EE

Published in the USA by
Routledge
a division of Routledge, Chapman and Hall, Inc.
29 West 35th Street, New York NY 10001

set in 10/12 Times
by Columns of Reading
and printed in Great Britain
by Richard Clays, Bungay, Suffolk

British Library Cataloguing in Publication Data
Ash, Juliet
Components of dress: design, manufacturing and image-making in the fashion industry.
1. Fashion industries & trades
I. Title II. Wright, Lee III. Series
687

ISBN 0–415–00647–3
ISBN 0–415–00648–1 Pbk

Library of Congress Cataloging in Publication Data
Components of dress: design, manufacturing, and image-making in the fashion industry/
edited by Juliet Ash and Lee Wright.
p. cm. – (A Comedia book.)
Bibliography: p.
Includes index.
1. Clothing trade – Great Britain. 2. Clothing and dress.
3. Fashion. I. Ash, Juliet, 1949– . II. Wright, Lee, 1957–
TT496.G7C66 1988 88–4388
687'.0941–dc19

To
Jesse and Annie (J.A.)
Audrey and Marshall (L.W.)

Contents

Section III Marketing Strategies

Section IV Class and Clothes

Section V Specifics of Gender

Section VI Manufacturing Images

Preface

The essays in this book are edited from theses written by third-year BA (Hons.) fashion-design students at Ravensbourne College of Art and Design over a three-year period (1984–6). As tutors we have worked closely with the students. However, the views expressed in the extracts are not necessarily compatible with our own.

Many of the student authors are now completing postgraduate studies in fashion or textile design. Others are working as professional designers abroad and in the UK.

Acknowledgements

Thanks to: the library staff at Ravensbourne College of Art and Design for their help, assistance, and ongoing support; Robert Prince and Lucy Chamberlain for typing; Bruce Robbins and Nigel Bradshaw; the Fashion Department staff at Ravensbourne for lunchtime laughs and their willingness to 'talk shop'; numerous museums and libraries, including The Victoria and Albert Museum, The London College of Fashion, The British Library, Brighton Museum and Art Gallery, The Costume Museum, Bath, The Museum of Mankind, London, The College of Distributive Trades, London, and The Imperial War Museum.

General introduction

Fashions may be transient, but the fashion industry itself embodies a diversity of cultural, social, and economic histories. As an industry it appears to live out the capitalist dream of constant and overflowing production supplying an insatiable consumer market; but how accurate is this impression? This book offers an insight into a post-industrial product created by an industrial society and investigates many aspects of the fashion industry.

The business of fashion has a huge earning capacity. Since the nineteenth century, when mechanization turned clothes-making into an industry, the business has expanded into a massive but divided conglomerate. The industry is renowned for its volatility and high-risk susceptibility to economic pressure, and market research and fashion forecasting are increasingly playing their part in an attempt to reduce the precarious nature of the business. Design as part of the production process can also help to offset many of these risks, yet in Britain designers are often given a low status within a company. The originality that is fostered and catered for in the art and design schools is not seen to be wholly pertinent to the home-based industry; abroad, however, the employment of a British-trained designer who is a product of the art and design school tradition is often the key to a company's success. Paradoxically there is an increase in applications to attend fashion-design degree courses just when the educational structure is being economically undermined. Design in general is becoming much more part of the popular consciousness in Britain today; and abroad, the Britishness of fashion can be an important selling point, indicating that a long-term investment programme is desirable in further education in this field.

Fashion appeals across class, gender, and racial divides and commands respect regardless of linguistic and/or cultural disparities. It is a cross-cultural form of communication and economic exchange, yet in their country of origin designers are regularly derided by managerial élites for being too adventurous and therefore commercially unsound. Ironically it is the imaginative nature of our trained designers which makes them eminently employable and sought after abroad. While design as a trained profession via education is forward thinking, there has been a commercial reluctance to tap its potential – although currently there does seem to be an increase in awareness of the importance of design on the part of home-based industries.

Fashion-design teaching in Britain is based on a system of art school training established in the nineteenth century. Changes have taken place as design has established itself as a profession in its own right, and since the 1960s the inclusion of academic and business courses running parallel with the practice has been considered integral to the modernizing of design teaching. On BA degree courses the practice of design is about 'doing' and 'making', combined with a small percentage of 'thinking' and 'articulating'. The latter can have many positive results for a practising designer, as this books aims to show. The 'usefulness' of history and theory in addition to studio practice is exemplified by the extracts. There are three major benefits which we have seen result from the 'academic' strand of design training: history and theory can act as a design approach; it can aid the understanding of the process with which designers are concerned in order to contextualize their product within the whole field of design; and it can help develop a sense of objectivity and criticism which is often lacking in the discipline itself. We have chosen to explore fashion in terms of its industrial context in order to address the broader issues which stylistic histories do not.

The term 'fashion' is all-encompassing in that it describes a totality made up of the garment, the method by which it is produced, and the meaning which the garment generates – that is, fashion as a form of visual literacy. These three components are part of a complex infrastructure and individually are worthy of much greater study than we can achieve in this book. But we hope nevertheless to provide some insights.

Fashion also depends on the ability to sell the product. How many times have you looked at the fashion pages of a magazine and seen fashions that you have never seen being worn in real life? Up-market mail order may be beginning to boom, but the high street is where fashion is reflected. The majority of people gauge what is fashionable both by what they see in glossy magazines and by what they see *en masse* in shops and being worn in the street – and rightly so, since 'fashion' images need to be seen as wearable permutations in order to register as a definite fashion in clothing. If the process by which that is achieved is complex, then questions such as

Is fashion a reflection of consumer taste?
Does it dictate that taste?
Does it set patterns of consumption?
Does the consumer really have any choice?

have no one simple answer.

The fashion industry is characterized by fragmentation and division but it still manages to create a 'glossy' mystique and successfully perpetuate an image of glamour. Up until the 1960s each season tended to be dominated by one or two images of how a man and a woman should look in order to be fashionable. Since then predominant style has become less visible, and there are now many different images which are presented on the catwalks of the world, in magazines and shop windows. Confusion is often experienced by the manufacturer and the consumer alike. Perhaps pluralism is just as intimidating and dictatorial as its predecessor? Despite the antagonism and the contentiousness fashion can generate, it is still one of the most personalized and accessible forms of self-expression. There are a number of studies which undertake a psychological exploration of clothes as representing attitudes of the wearer, yet few ask questions relating *meaning* to *production*.

In the nineteenth century mechanization and mass production changed the nature of fashion and laid the basis of the industry as we know it today. *Haute couture* also began at this time, high fashion for the élite, at the other end of the spectrum from mass production. *Haute couture* still brings a kind of hand-made status to the industry. In fact, it is this small sector which often has the ability to experiment, since limited production and high cost are usually offset by a cost-effective ready-to-wear collection and a financially secure clientele for the more exclusive designs. Design history and theory facilitate the study of fashion on many levels, and this book touches a variety of examples of this. The basis for designing clothing is that the end result has to meet certain criteria that are not necessarily set by the designer. Design education concerns itself as much with the profession of design as with designing and making.

But what is designing fashion all about? It is less about invention than about innovation, using established forms such as the dress, the suit, and the sweater as the basis for reworking. Whatever the end result, the one principle is that it must house the body. Fashion constantly sets up a relationship between itself and the form underneath it. In industrial design in general, the notion that the exterior covering has to follow the dynamic character of the form, be it the engine in a car or the human physique in clothes, has been a major design approach in the twentieth century. But to say that this is the only influence on the visual shape of the object is to oversimplify the practice of design. The functionalist argument that all objects are, or should be, designed from the inside out does not totally explain the directions that fashions take. The fashion industry has the stylistic ability to change and adapt very quickly, and often these changes are brought about, we would argue, by external factors such as the social and economic climate.

The complexity of the industry cannot be understood either by the utilitarian argument that all fashion has to be functional or by the theory that all clothing is expressive of a degree of affluence. Attempts to discuss and theorize fashion and design have been made in journals such as *Block, New Society,* and *Women's Review*. The dialogue continues, mostly in art and design schools, and particularly in the history and theory departments which work alongside the studios. Students complete a course in historical and theoretical studies, culminating in the writing of a dissertation. It is from these final-year theses that this book is compiled.

The process of researching and writing is not trouble free for design students. Much moaning and groaning goes on, and Biro stains are smudged over prototype clothes or *toiles*. Every year complaints become part of the ritual:

GENERAL INTRODUCTION

'How can I write when I have thirty finished fashion illustration sheets to complete by tomorrow?'

'Honestly, I *am* dyslexic. Can I make a tape instead?'

'I just like designing in leather. What's the point of writing about it?'

'I only read magazines. I don't read books.'

Each year the same agonizing occurs, interspersed with peaks of excitement as the results of research generate new lines of questioning:

'Men wore high heels and frock-coats in the eighteenth century, prior to the French Revolution. What did this mean to 1984's men in skirts?'

'Roland Barthes must have been mad to think fashion magazines were to do with a linguistic theory. They're all made up, so they are fiction!'

'Did women at women's liberation conferences in the 1970s really take *all* their clothes off? I thought they were into covering up in dungerees, like men.'

For many students the lengthy piece of academic work results in new understanding, and for them fashion becomes imbued with new meaning. As future employees or employers, fashion-design students need to be aware of some of the arguments and approaches connected to the industry. The discipline of design history has facilitated students in looking at the history of the fashion industry and the way it operates.

For us as tutors, the results of an integrated system connecting theory to practice are encouraging. Each year there are a whole new set of subjects and ideas, and we seem to be pleasantly surprised, if not overwhelmed, at the potential in the students' work. The style may be weak, many of the links in the argument may have been left out, but mostly it is fresh material and original. The complex nature of the industry provides an added challenge and makes teaching all the more exciting.

The variety of primary sources students have to have recourse to range from anthropological museums to business archives and newspaper libraries. This in itself shows up the lack of contemporary material about fashion. There are regular costume histories which run through a chronology of style; and there are the high-flying theoretical books which have to be decoded (for example, Barthes's *Fashion System*). But most of these books do not relate history and theory to the practice of design. Recently a few books (for example, Elizabeth Wilson's *Adorned in Dreams*) have been published which provide socioeconomic source and back-up material for students, but they still do not apply the history of ideas to the *making* of clothes.

The six sections that follow each take a different aspect of the industry, past and present. We have not attempted to outline fashion events chronologically. Such a history can be misleading in that it implies that each event is a consequence of and a development from its predecessor. Instead, the thematically organized sections each examine a major issue relating to the fashion industry. Each section presents a range of topics which supply various views and insights into the fashion industry, in order to give an overall impression of how the fashion garment is produced and reproduced. Of course, covering so much material means that, while many points and questions are raised, detailed substantiation is often lacking. However, our intention is to provide a stimulus for students and designers to inquire further into their chosen profession and to use such research to aid the practice of design.

The extracts selected for inclusion are not necessarily the best examples of students' written work, but they are pieces which have attempted to come to terms with significant ideas which clarify particular fashion practices. We have linked the extracts with an introduction to each section in order to give some idea of the continuity of thought, and in some cases to raise more general issues that the essays themselves touch upon. Although it may seem clichéd to start by examining historical developments of the fashion trade, there are certain aspects of the industry which have been regularly glossed over by both critics of the industry and those who work within it.

Section I begins with a look at the production of clothes which very quickly demystifies the glamorous world of fashion. To understand the present production structure of the industry, which is divided between 'sweated' and factory manufacture, it is necessary to return to the Industrial Revolution and the roots of the industry as an organized system. Chapter 1 asks the question, why is such an outmoded form of industrial practice such as sweated labour still in existence in clothes production, when most industries are being transformed by technology? As the second essay in this section (Chapter 2) illustrates, the practices of manufacture are contextualized by the means of production. It examines the nineteenth-century factory system in northern textile towns in relation to the workers and the traditions of the work environment. The other two essays in this section consider two strands of clothing production in greater geographical detail. Chapter 3 is a piece of oral and social history concerning the making, distribution, and wearing of clothes in a part of rural England between the wars, and Chapter 4 inquires into the reasons for the emergence of cycles of immigrant labour in the East End of London. From this first section it is evident that there are close links between the textile and clothing industries.

Section II examines different aspects of the commodities from which fashion is made. This concentrates on the nature of fabrics, without which fashion could not exist. The initial essay (Chapter 5) presents an analysis of printed textiles and discusses the connections of surface decoration with both the two-dimensionality of painting and the three-dimensionality of sculpture. This is followed by specific inquiries into the nature and use of tartan (Chapter 7) and leather (Chapter 8), and how both have traditions of functionality which have been transformed, via use and meaning, in the twentieth century to convey ideas of rebellion or fetishism amongst particular groups in society. There are fashions in fabrics be it leather or tartan and/or fabric-making such as knit. These preferences parallel fashions in clothing. Sometimes the two are so closely interwoven that the type of garment is dominated by the fabric from which it has been made. The other essay in this section (Chapter 6) illustrates the vicissitudes to which fabric is subject in terms of supplying the demands of the fashion industry.

If production is concerned with the making of fashion, then the retail trade and consumption complete the cycle. These areas of the fashion industry are often simply equated with business studies and merchandising and not treated as part of broader cultural issues. The essays in Section III investigate a number of aspects of marketing. The first essay (Chapter 9) studies design policy in a remarkably successful high-street chain store, whilst the second (Chapter 10) examines window display as a method of selling the product. After all, fashion is a product which, like any other, needs to be profitable in order to justify its production. The third essay (Chapter 11) takes a broader look at one type of inspiration for fashion designing and considers how ethnicity is exploited in its marketing.

Ways of marketing are not the end of the story, since the consumer exercises a certain amount of control over the industry in various ways. Consumption is not merely about exchanging money for goods; it is also about choices that are made, and the way people select, according not only to the money available to them but also to their own social grouping. Section IV looks at the way two privileged strata of society, at differing historical periods, have made decisions as to their clothes buying, based on traditional values and a 'safe' approach to the new. These essays (Chapters 12 and 13) concern women as consumers, in contrast to the following section, which deals with another group of fashion consumers who have begun to reassert themselves after years of slumber.

Traditionally, men have been associated with rather regimented forms of dress, while the fashion industry has concentrated on the female as the gender which dresses up. In Section V four areas of masculinity are investigated, in order to review gender-specific design. Since this is the more uncharted area of product design, segregated by the sex of the consumer, many questions are raised about dress and undress; and the essays (Chapters 14 to 17) deal with the subject in a variety of ways which cannot necessarily be transferred to the female counterpart. This approach shows that history and theory can be appropriated to specific areas of study in order to challenge stereotyped 'readings'.

Finally, the translation (and often the transformation) of fashion ideas is discussed in terms of media imagery. Fashion, more than any other art or design discipline, is constantly placed under public scrutiny in the form of exposure in hundreds of magazines and newspapers throughout the world. Section VI looks at the aesthetics, techniques, and content of the printed visual imagery of fashion with which we are constantly bombarded (Chapters 18, 19 and 20).

As a whole, the extracts are indicative of a typology of subjects that have interested students in the 1980s, and have been selected and edited in such a way as to link components which make up the fashion industry. While they concentrate on fashion, the areas explored are also relevant to other industries linked with design, mass production, consumption, and communication. This book will be of interest not only to fashion students hunting around for ideas but also to their future employers, whether designers, retail traders, fashion editors, or other fashion-orientated business people who may be interested in the wider concerns of students and the effects this research and conceptualization of dress could have on their design practice. Those involved in school education will also find in the book aspects relevant to secondary teaching as design becomes increasingly important in the school curriculum. It may even be of interest to those of us who are sceptical of an industry which is so volatile and yet entraps us in its magic – an industry which is such a rat race and yet constantly generates a creativity powerful enough to transform our lives on a daily basis.

SECTION I
Urban and Rural Practices

INTRODUCTION

A transition from a rural to an urban community in the mill towns of England continued into the first half of the nineteenth century. It was initially a gradual influx of people moving from the country into the towns, becoming more rapid from the 1830s onwards. The urban areas expanded to accommodate the increased population, swallowing up large areas of the countryside. The organization of labour was the basis for this expansion, and the rate of growth was consistent with the variety of means to earn a living. Lancashire, for example, as the centre of manufacturing, experienced more rapid expansion than many other areas. By the beginning of the twentieth century, society had undergone many changes socially and economically; and while the process of urbanization was not complete, the modern demographic pattern had been established.

The mass manufacture of clothing depended on an industrialized process, and it was only after the 1870s, when the sewing-machine revolutionized the method of manufacture, that a ready-to-wear industry took off. However, the industry did not undergo instantaneous change from the production of clothing by hand to its production by machine. In some sections of the industry hand production has continued to exist side by side with industrial methods.

There are many different types of manufacture, ranging from extensive mechanization to small-factory and home-based production. Since the eighteenth century, profit margins have usually been at the expense of a cheap labour force, whether at home or abroad. When the exploitative nature of the rag trade has come under trade-union pressure from its home work-force, many companies have turned to cheap production abroad in places such as Taiwan, Hong Kong, and Portugal. This problem did not arise in rural areas, since the crafts person served the immediate

Figure 1 Lees and Wrigley Spinning Mill, Oldham, *c.* 1890 (Oldham Chronicle). The nineteenth-century factory was a new phenomenon in both scale and mass production. The collective workplace was a product of the Industrial Revolution and textile factories dominated the landscape in parts of North England. Never before had the difference between rural culture and urbanism been so obvious. As a percentage of the population moved to work in the factory system, rural practices began to be undermined and in some cases lost forever.

community, using one or two local apprentices. This was very much a part of the social fabric of the village.

Community relations are not necessarily alien to urban production. The East End of London is one example of an inner-city area which has had associations with garments and textiles since the eighteenth century. One of the reasons for small companies utilizing this region for cheap production is its close proximity to wholesale and retail markets in the West End of London.

The essays in this section deal with some of the above issues, both past and present. The main concern is with the social relations of production in the textile and garment trades.

As consumers, we are often unaware of the structure of the industry and the methods of production which, although changing rapidly in some areas of production (for example, computer-aided design), retain in other sectors of the industry outmoded forms reminiscent of the worst conditions of the nineteenth century. This first section illustrates the many factors which have

Figure 2 Brooks Factory, East End of London (Photo: Michael Ann Mullen/ Format). Garment production often relies on a wealth of small businesses located near the retailing centre of fashion. Fashion changes allow for a rapid turnover of capital and cheap production costs. This is based on a long tradition of low wages which perpetuate the 'sweatshop' syndrome.

affected and continue to affect the fashion industry. It concentrates on the manufacturing process of garments and textiles, and the four essays provide an insight into the fragmented nature of the industry.

1

1884 in 1984: the progress of clothing sweat-shops

ALEX BUGLASS

Everyone complains about the price and quality of clothes in the shops today, but do they ever stop to think about the crowded rooms filled with sewing-machines where groups of women have to produce large quantities of finished garments just to earn a subsistence wage? Consumers of clothes are often unaware of, or choose to ignore, the fact that the dressmaker's struggles to survive continue. Perhaps we are afraid that a reform in the clothing industry would mean a large increase in the price of clothes. But what is more important – shirts and skirts, or flesh and bones?

With the aid of writers and reformers, many of the injustices of the Victorian era were brought to public notice and dealt with. The plight of dressmakers joined that of chimney-sweeps, mine and factory workers, most of whom saw a great improvement in both conditions and wages. Unfortunately, the stratification of the clothing industry has ensured that clothing workers' conditions of employment and wages have changed little during the twentieth century.

Needlework was a subject nearly all Victorian girls were taught, yet there was a vast difference between the pleasurable sewing enjoyed by the upper classes and the sweated labour of the needlewomen and labouring dressmakers. Sewing was one of the few routes open for women to earn money in those days, and the vast number of dressmakers was due largely to a new market for clothes. In earlier days fashion had been a luxury for the rich. Fabrics were expensive, and communications were slow, therefore styles were slow to change. Improved communications, especially on the railways, meant greater mobility. New printing techniques and cheaper paper resulted in the growth of women's magazines, which devoted a lot of space to fashion. Clothes production became an organized industry, and the demand for clothes was greater than it had ever been.

Utilitarian economic ideas were a major part of Victorian philosophy. Competition forced prices down, but this meant lower wages; and, without the benefits of a Welfare State or a trade union, the seamstress had either to accept the meagre wage she was offered or to go without. The rise in population meant that there was a plentiful supply of cheap labour. There were over 20,000 dressmakers in London alone in 1850. Job opportunities for women at this time were very limited because of the belief that women should stay in the home. The imbalance of the sexes increased throughout the period, until the turn of the century when there was a surplus of over a million women; despite this fact, no attempt was made to give women jobs which had previously been undertaken by men. The females of this era could do only women's work in the form of domestic service, dressmaking, or teaching. Women at this time were very much second-class citizens, and their grievances often went unheeded. Today, in the majority of sweat-shops, Asian and Cypriot women are to be found working. Their situation is very similar to that of Victorian women.

The introduction of the sewing-machine and the rise in working-class living standards in the late nineteenth century meant that the clothing industry expanded to accommodate the growth in the market. Production in factories developed alongside work carried out by individual women in the home. London had long been a centre for producing clothes for the rich and now became the headquarters of the mass production industry. This conversion was possible in London for two main reasons: firstly, because textile and garment manufacturers had access to the largest and, in design terms, the most dynamic market in the country; secondly, and just as important, there was a constant supply of cheap labour provided by successive waves of immigration. These two factors are just as relevant today.

Initially production in small firms concentrated on menswear because the early machines were more suited to it, and slower fashion fluctuations meant longer runs. Factories began to change their method of production to increase efficiency. This was achieved by the introduction of section work, which meant that each machinist worked on one part of a garment as opposed to making a complete outfit. The result was that garment workers became very fast in their job, but with this increased efficiency came the tedium of doing the same section day after day.

With increased investment in the mid- to late nineteenth century some factory owners, such as Sir Titus Salt, could afford to move out of the cities (that is, Bradford) and set up more hygienic and better-built factories in the countryside. Based on socialist utopian ideas, the New Model Factories – such as Saltaire (wool manufacturing) in Yorkshire, and Clarks shoes in Somerset – could provide their employees with improved conditions.

However, a London location remained a distinct advantage for smaller firms. These small firms were, and still are, reliant on their ability to turn out orders extremely quickly, thus enabling wholesalers to keep their stock low; this in turn reduced their vulnerability to changing fashion trends, and enabled them to respond rapidly to such changes.

'Sweating' is a well-known feature of the clothing industry, so much so that it is regarded as inevitable. The reason why the conditions are so poor, and why the people work in unpleasant surroundings, is that production is planned and governed by the invisible hand of the market and by the very visible hand of retail, wholesale, and manufacturing firms, operating with the aim of making substantial profits.

Some large retailers choose to move clothing production from one region to another in order to obtain a more compliant and low-paid work-force. Retailers and wholesalers often place orders with a large number of independent firms and frequently shift their orders from firm to firm. In this way they can obtain the lowest production prices and the greatest flexibility with the least risk to themselves. The industry is labour intensive, and much of the production is in relatively small batches. These economic aspects of the industry mean that capital involved in clothing production is easily mobilized and transferred between enterprises and geographical areas. The clothing market is especially vulnerable because it has to operate by serving the short-term fluctuations of the fashion market. The production which remains in London survives by coping with sudden changes in fashion and shortages of particular lines. In short, it is a 'perfect' market, but this 'perfection' works to the benefit of large wholesalers while the 'sweated' garment worker may be employed and laid off in rapid succession.

The image of the industry is that of back-street sweat-shops, poor working conditions, the production of low-quality garments, outdated production technology, and inefficient production techniques. There is a low level of capital investment and an abundance of sharp practices. It has the ingredients of an 'easy in, easy out' industry which attracts the fly-by-night entrepreneur who will liquidate a business to keep ahead of creditors and start up again under a new name. This image also deters retail majors from placing orders, and banks from lending money. The industry is changing, but at present much of it would be classed as technologically backward. London is worse off than other areas for several reasons. Many of the capital's clothing firms have very poor working environments. Modern premises are rare, and the age and design of older buildings are not conducive to producing pleasant, let alone efficient, working conditions. In many cases there is little or no regard paid by the owner to the conditions. Crowding is commonplace, and toilet facilities are often primitive. As shown by the numerous fatal fires in such places, statutory regulations are often ignored.

The clothing market is a volatile one in the sense that fashions are constantly changing. Between the end of the Second World War and the mid-to-late 1960s, fashion at any one time was dominated by several items of clothing which were intended to enhance a particular image; thus, for example, men's fashion in the mid-1950s was dominated by specific styles of trousers, shoes, shorts, and jackets. Some fashions were casual and others formal. But they would generally remain fashionable through one or two seasons. However, since the 1960s both up- and down-market fashion changes have speeded up, and producers now search for quick trends that appear and disappear over a short space of time; the London industry specializes in those fads. But this rapid turnover of type of product in the 'sweated' London industry is not accompanied by new machinery or higher wages.

A very different approach is adopted by such shops as Next and Benetton. Here they choose a well-defined social group and sell them a 'look', a set of co-ordinated garments as a sort of Habitat of the clothing industry. With this plan, the retailers are able to insulate themselves to some degree against each and every fluctuation in fashion and instead carve out their own part of the fashionable market. Hand in hand with these trends is a general tendency towards better-quality garments. The increased competition that companies such as Next and Benetton provide has serious implications for London production, which has a reputation for poor-quality garments.

What is happening in the industry world-wide obviously reflects on our own manufacturing sector. Both Western Europe and the United States have managed to stay ahead of the United Kingdom through vast investments in manufacturing technology. On the other hand, production in the Third World is based on low wages, although more recently these countries too have been investing in the latest machinery, thus competing with the UK in terms of both low wages and equal or superior production methods. So domestic producers are being squeezed by both the higher- and the lower-wage countries. The outcome of this will be either more work for lower wages (thus resulting in worse sweat-shop conditions) or redundancies and closures of these small factories in preference to the Next and Benetton type of production.

Until recently, technological change in the international clothing industry has been slow. Production was centred around the technology of the sewing-machine, which had changed very little since the mid-nineteenth century. Recently, however, the microprocessor has enabled cheap computing power to be applied to all stages of production. Computer-aided design (CAD) is being introduced into the design stage, the fitting of the pattern to the material (laying), and the adjustment of patterns to cut garments of different sizes (grading). Even the traditional skills of the cutter are becoming computer aided. However, since small 'sweated' firms survive on little or no capital, this new advanced machinery has little hope of reaching them.

The fashion industry is and always has been one of extremes. It is hard to believe that the luxurious showroom of a fashion house could have any connection with the dark and crowded conditions of an East End sweat-shop. For too long now we have been happy to buy cheap clothing with no thought as to its origins.

2

Factory production in the cotton industry in the eighteenth and nineteenth centuries

MANDY LAWRENCE

Cotton was at the forefront of the new wave in industry in the eighteenth century, and its expansion was to become a major factor in the creation and development of the English textiles industry.

The woollen industry was England's greatest and oldest trade after agriculture. Cotton was of minor importance until the end of the eighteenth century. Before this time the cotton industry was for the most part what economic historians call domestic. That is, it had been carried on not in factories or large workshops, but chiefly in the cottages which were the workers' homes. It was customary in many homes, especially if there was more than one weaver, for the women and children to do the carding and cleaning, and for some of the rovings to be sent out to be spun. This would be performed by neighbouring spinsters, and by wives and daughters of craftsmen. Some of these were full-time spinners who depended upon this work as their main source of income; others took in spinning as a by-employment.

The machines which revolutionized the industry created the factory, as they facilitated mass production. At the same time as these advances took place in the design of the textile machines themselves, a new power source was developed: the stationary steam-engine. The first rotative-beam engine was installed in a cotton mill in 1785. This meant that textile mills no longer had to be sited in rural areas within easy access of water, but allowed for the establishment of factories in urban areas, from which sprang the growth of the mill town. The introduction in 1764 of a mechanized spinning machine, the spinning-jenny, implemented the decline of spinning as a domestic industry. Although this brought much anguish among spinners, not all attitudes to the jenny were as hostile as might be supposed. At first there had been an outcry that the jennies would take bread from the poor. But many of the spinners were wives and children of weavers who saw that the increase in output would be to their benefit, and were only too glad to adopt the new machines, in a new environment. Those who suffered were the old and people who could not afford these machines. With the introduction of machines for carding and roving yet another line of production was moved into the factory workshop.

With the introduction of steam power in the nineteenth century new and more efficient machines came into the industry, and cotton became cheap and plentiful. By the beginning of the nineteenth century, cotton had seized an 80 per cent share of the textile market, with wool trailing behind at only 16 per cent. Prior to the nineteenth century there had been an Act (1720) to limit the production of cotton and keep prices high, since it was considered a threat to England's indigenous wool and silk trades. In the nineteenth century for the first time cotton or fustian (made with a linen warp and a cotton weft) became affordable to the middle and working classes. Fustian was particularly popular because it was very warm and hard wearing. Both cotton and fustian were considerably cheaper than wool and linen and, being easier to launder, brought about a major improvement in hygiene.

With the increasing popularity of cotton came the expansion of the cotton industry and the consequent increase in profits for the mill owners. Britain's industrial might, of which textiles were a part, created and supported the empire, as thousands of fortunes were made by the entrepreneurs.

This era of great expansion also saw a shift in concept from machines as tools, as a service, to the other way around. The machine became the new master, which everything and everyone had to serve, and it very quickly showed itself to be as exacting and cruel a master as the worst of the human kind. In this case the machine changed not only the nature of the product but the lives of the people involved in textile production. But there was a shift also in Britain's textile trade, which had been dominated by wool. By destroying India's indigenous cotton industry and expanding its own in the nineteenth century, Britain sowed the seeds of part of its own industrial decline in the twentieth century. It was because of the British development of the cotton industry at India's expense that Gandhi was later, in the 1920s, to boycott British textiles in order to build up India's own production and export of cotton manufactured garments. The decline in textile and garment production in the UK has continued into the 1980s.

3

The life of a Suffolk horseman (his work and dress), 1910–45

EMMA MITSON

The life of rural Suffolk during the period of 1910–45 saw many changes. Some were for the better, with a higher standard of living for all; other changes, the old horsemen of Suffolk will tell you, were for the worse. Although Suffolk, always unspoilt and unprogressive, was one of the last areas to become mechanized, inevitably the tractor and the combine harvester came. Before the days of the tractor, the horse was the principal source of motor power on the farm. Without this power, the raising of crops on any scale would not have been possible. The horse also solved the transport problems of the countryside and was the centre of the economic life of the region. The care of horses on a farm was a job of great importance. Large farms would employ a head horseman, second horseman, and so on, each with an allotted number of horses to care for.

The farmers and horsemen of Suffolk suffered unemployment during the Depression of the 1930s, which led to an exodus from the land. The demands of the war machine found jobs for many, and the Second World War saw the men of Suffolk in the armed forces, returning after it to a new agricultural world in which the prosperity of farmers was founded on the efficient production of food with machines instead of manpower. The rhythm of the old life, with its economic hardship, its insularity, its fixed social station, gave way to a standard of material prosperity unknown to earlier generations. Old rural communities became infiltrated by business commuters from the towns, and a breakdown of the previous social rigidity ensued.

Living in a Suffolk village offers a great opportunity to learn about the life of the agricultural worker. Memories are recalled, and an old people's tea-party brings back a fund of information. Kersey is an old weaving village nestling between two hills. The church dominates the top of one hill, the ruins of a priory the other; across the middle of the street a tributary of the River Brett bubbles. The weavers' cottages are still there, though the Kersey cloth for which the village was famous (and which is mentioned by Shakespeare) is no longer made. The old wattle-and-plaster houses are mostly owned by newcomers. The cobbled street has been tarmacked, and the baker, cobbler, and butcher are no longer there. But the close-knit family tradition persists, and memories are fiercely guarded. The horseman is no more, but he should not be forgotten; what follows is a reflection of the working life of a Suffolk horseman spanning those years of change and recalling memories of education, social life, clothes, and fashion.

The story of the Suffolk horseman starts prior to the First World War, the time of the invention of the motor car. With country folk you can look back through history. In these small Suffolk villages people are steeped in tradition, and their memories flow back through their parents to their grandparents; they are full of tales, mostly of the good times, but times passed down by word of mouth and well remembered. The First World War had a dramatic effect on the use of horses. It is difficult to imagine, with our present-day sophisticated weapons of war, that in 1914 horses were needed both to pull the gun-limbers and for ordinary transport on the Western Front. Horses were being killed almost as quickly as men, and by 1917 the shortage of horses got so acute that the military authorities began to commandeer horses wherever they could find them. An Ipswich man recalls seeing a horse on an Ipswich street being unhitched from its cart by a party of soldiers: shades of the press-gang! The cart was left on the side of the road for the owner to get home as best he could. The horses that did survive the war were often so shell-shocked that they were no longer a great deal of use on a farm; the noise of a gun would terrify them, as indeed did the sight of a soldier in uniform.

The war helped to escalate change in the countryside, and many of the big estates were broken up. Steam power, which had been introduced in farms before the war but with little impact, now started to revolutionize farming. Life in many places took on a different character to that of the previous two generations. Once the war was over, everyone assumed life would go back to its steady rhythm; but things were never to be the same again, even in East Anglia.

East Anglia was not noted for the benevolence of farmer to farmworker. The hours were long, though the lot of the horseman (who was regarded with some respect) was considered much better than that of the farm labourers. The horseman's was a sought-after position. Bertie Baalham, now nearing his eighties, recalls how as a young lad he worked on Saturdays as a yard boy for the Partridge family in Kersey. He swilled and swept the yard, cleared up, and did any work required around the back of the house and yard. At 11 he left school and was taken on by the Partridge family as back-house boy and to help the stockmen. He carted the water and straw to feed the cattle. The water cart held 160 gallons, and he refilled it from ponds and streams, sometimes as far away as Lindsey (some miles from the farm). Later he became back-up boy. This meant that if one of the horsemen was off sick, he took his place and became responsible for the horses in his care. One assumes that he did a satisfactory job, for eventually he became the junior horseman of nine horses, and at this point in his working life his prospects looked good.

Horsemen were paid a fixed weekly wage, and also had certain perquisites which put them higher in the farm organization, in fact on a level with the stockman or the shepherd. Ernie Spraggons (another Kersey horseman) started work at 13 years of age scaring rooks, for which he was paid half-a-crown a week. His wage as a horseman was fifteen shillings before the Second World War. There were ways of earning extra money. One of these was stone picking. The fields of East Anglia are littered with large flint stones. Not only did the farmers believe their crops would grow better without the stones, they also needed the stones to make paths and to use in the upkeep of roads. Ernie Spraggons relied on stone picking to provide clothes for his family. In many families the wives and children did it, for money was short. Pigeon-shooting earned extra money, and rabbit-catching could provide extra food for the family.

As the effects of the Depression were felt, many farmers were unable to cope with the economic gloom which had overshadowed them. While they sold land and equipment, it was the selling of the farm horses that caused great grief to the horseman. Bertie

Baalham recalls the sadness with which he parted with old favourites and even today he remembers the names of two of them, Belle and Prince, and part of a lifetime spent together.

Before the First World War, dress was rigidly limited both by the class and by the occupation of the wearer. It was the mark of a more stratified and more authoritarian society. The general expectation was that a man should dress according to his station. Each tradesman had his particular dress and usually he was proud of it. A retired blacksmith has said how in his younger days he would walk into the street in his leather apron, go to the pub or post office still wearing it; he was proud of it. On the farm, too, there was a tremendous pride in occupational dress. The horseman in Suffolk was paid more than the 'day man', so his dressing well merely confirmed the hierarchical structure of the village at that time.

It is said that the sleeved waistcoat as we know it started in Suffolk. The farmers made men work with their jackets off all the winter, because they would have to work harder to keep warm. But the women noticed that the white inner sleeves of the jacket, if turned inside out, looked as if the men had shirt sleeves and a jersey pullover on. So the men turned their jackets inside out so that the white inner lining of the sleeves was showing, and from the farmer's house it looked as though the men were working without their jackets. Farmworkers recall the days when they used to turn their jackets inside out when they got to work and at dinner time turn them round in case the farmer turned up.

The tailor travelled round the villages in his area. Frank Whynes, who was born in 1891 and lived and worked from Stowmarket, has described the horseman's dress in detail:

> I had to make the sleeved 'weskit' (waistcoat) according to the customer's order. Nearly all of them had a hare-pocket on the inside, on the left. If you were wearing one of these weskits you could double up an old hare and place it inside and button up; no one knew anything about it. Some men wanted this pocket extended right round the back of the weskit. If they wanted the pocket lengthened this way, I had to fix eight buttons to fasten the top of the pocket to the lining of the weskit; if I didn't do this, when the pocket was full it would sag below the bottom of the weskit and give the game away. Occasionally a man asked me to sew in a gun-loop high up on

the left side of the weskit – inside of course. This loop held the barrel of the gun while the butt rested on the bottom of the hare-pocket. With a gun-loop and one of these pockets, the horseman could walk about as though he had no concern but to get a proper draw on his clay pipe.

For walking out and on Sundays, the horseman had a cord jacket and cord trousers. The trousers were not ordinary trousers; they were 'whole-falls', that is, trousers with a flap that let down in front like a sailor's. They were also bell-bottomed, with a 16-inch knee and a 22-inch bottom. The outside of the trouser leg was trimmed with steel-faced horseshoe buttons. Some of the more dress-conscious horsemen ordered a special kind of trimming on the leg – an inter or gusset of black velvet running from the bottom of the trousers, on the outside, and tapering to a point somewhere near the knee. Four or five horseshoe buttons were sewn to this gusset as an extra decoration.

The material that was used for clothes at that time was often staple tweed. It was hard-wearing and could last fourteen to fifteen years. Overcoats were made in Melton, which is a thick, very tightly woven woollen cloth. The weave is so tight that water will never get through. This bears no resemblance to the less durable Melton cloth of today. In earlier years an overcoat could be passed from father to son and to his son before it was worn out. In Kersey, Bertie Baalham remembers that two tailor's men from Ipswich named Walsh and Bulberry would call round at all the houses to see what clothing was required. After your order for a Sunday best suit was delivered they came back once a fortnight to collect the money. No one paid outright because they didn't have the money. When the suit began to get 'messed up' it became a man's working suit.

When Bertie Baalham first started to work, before he became a horseman, he had only one suit, and so on Sundays it had to be pressed and cleaned to wear for church. Shoes were an important item because they had to be weatherproof. He recalls, 'Shoes lasted a long time. You wore hob-nailed boots with heel irons and toe tipe to make them last!' Busketts, which were like leather gaiters, were worn over the lower legs, meeting the boots, and these kept the legs clean and dry. Later, the wellington boot was used in winter for digging ditches or drains, but not for ploughing.

When farmworkers in Suffolk made one of their rare visits to the town it was usually after they had had their harvest wages, and they did their buying for a year. Arthur Price, a Stowmarket clothier, described them coming to the shop in his father's time, about 1920:

> These old country bo's would come into the shop to buy their ware, their outfit. And Father always kept underneath the counter in those days a 4½-gallon tub of beer (it was bought locally for about three farthings a pint); and before they started buying they'd say to one of us, 'What about it?' We knew what they meant. That would mean a glass of beer before they started business. Well, when they'd finished that, they'd look around – time was no factor; they were in town for a bit of a spree as far as their money allowed them – and they'd say, 'Well, might as well wet the other eye, guv'nor.' This meant, of course, they wanted another glass of beer before they started buying. Well, we gave them another glass; they finished it, and eventually got down to business. Probably they wanted to buy a pair of cord trousers. But they would want to pay for these cord trousers before they bargained for another item. They wouldn't really trust us to count up a lot of items. They couldn't count very well themselves, and they bought very cautiously. But they'd pay for the trousers and then ask for a pair of hob-nailed boots. They'd then buy a shirt and pay for each item individually right through the outfit. They could buy the lot out of their gold sovereign and still have some change.

Ernie Spraggons still possesses one of his working shirts. It was called an Oxford shirt, and round Kersey all farmworkers bought them. They were made of thick cotton and lined across the shoulders and chest so they were warm and windproof. With the extra money from harvest and stone picking, Ernie Spraggons was able to send his wife and daughter to Ipswich, where they would buy their clothes for the year. Many families made all their own clothes, and children's dresses would be made from the best parts that remained from father's shirt.

In looking at rural clothing production it becomes clear that alongside the mass production of garments and the expansion in consumption of clothes of the late nineteenth and early twentieth centuries, there existed the older more traditional production of one-off garments. Village communities were still reliant on the local tailor and/or dressmaker for their clothes, or made them themselves. In urban areas the hand-made production of clothes was subsidiary although it continued parallel to industrialization.

It has been my privilege to have interviewed and recorded the recollections of Ernie Spraggons and Bertie Baalham, horsemen of Kersey, who lived and worked through the period concerned in my research. I am also indebted to the people of Kersey who helped me in this project.

4

The rag trade: a study of the Jewish inhabitants of the East End of London in the clothing industry

HENRIETTA GARNET

Between 1870 and 1914 Britain experienced perhaps the greatest flood of immigrants until the Commonwealth influx after 1945. Jews arrived by the thousand, fleeing from Eastern Europe. A large number went to the United States, but approximately 200,000 came to Britain. A clamour arose, built on fear rather than fact, that the Jews were robbing the British of their jobs and homes. The Jewish immigrants became scapegoats for all ills. This prejudice has persisted to plague the most recent mass migration of the Indo-Pakistanis in the 1960s. One of the largest concentrations of Indo-Pakistanis in Britain is in the East End of London and surrounding areas. They too are subjected to ridicule and prejudice, for their dress, way of life, and foreign tongue, but mainly for their colour. It is interesting to see how this immigrant group has commandeered the once solely Jewish rag trade and is repeating the Jewish pattern of development.

For hundreds of years the Jews had been associated with the clothing industry. Traces as far back as 1753 are found of Jewish merchants selling second-hand rags in Petticoat Lane and in the markets surrounding Houndsditch. Up until the 1830s tailor-made garments were strictly a privilege of the more affluent classes. Those unable to afford the luxury of bespoke garments dressed in home-made, second-hand, or 'clobbered' clothes, the latter being hand-me-downs or bought from a pedlar or stall. In 1830 Jewish firms such as E. Moses and H. Hyam took note of the more clothes-conscious generation of working-class people and set up several 'slop shops'. They took advantage of the availability of cloth and made suits from cheap 'cabbage' cloth or used old suits to make children's clothes and sold them at affordable prices. In 1851 a contributory factor to the viability of the sale of cheap clothing was the invention by Isaac Singer of a practical sewing-machine and John Barran's band-knife cutting machine. *The Economist* commented in June 1851 that the sewing-machine would 'extinguish the race of tailors', but in fact it broadened the market, creating employment for many unskilled workers. In a survey of the industry in 1888, Charles Booth declared that there were 901 Jewish tailoring shops in and around Whitechapel, 685 of which employed approximately ten workers.

In most countries the capital city is the leading centre for the clothing trade, and this is certainly true of London, but the next and most important town involved in the development of the clothing industry was Leeds, in the production of wholesale clothing. A factory opened by John Barran (inventor of the band-knife cutting machine) in Alfred Street, Leeds, in 1856 was equipped with three sewing-machines and employed six cutters and twenty workers. His choice of Leeds as against anywhere else was probably due to the availability of cloth produced in towns close at hand. It may also have been because of the recently developed railway and water transport and the availability of coal to power the factories.

By this time, too, the small western or Soho colony of Jews, mainly from Germany, were being recognized by the Textile and Garment Union as important members of the tailoring trade.

With the arrival of the machines came a new idea of 'many men, one garment', with individuals being responsible for piece-work – for example, making sleeves or pockets. Developments after the middle of the nineteenth century were fast, and within forty years the wholesale trade was well established for the manufacture of men's clothing. Alongside the massive increase in production and the creation of new jobs, there was a further influx of 100,000 immigrants from the pogroms of Eastern Europe.

In 1900 the clothing trade was categorized into four groups:

(1) (a) Tailoring – retail bespoke.
(b) Tailoring – wholesale, including men's wholesale tailoring, the manufacture of men's and boys' suits, coats, and outer clothing, generally bespoke or ready-to-wear.

(2) The mantle trade – the manufacture of women's and girls' costumes, coats, skirts, and outer clothing, generally of heavy material.
(3) Light dressmaking – underclothing and outer garments of light fabric for women.
(4) Shirts – manufacture of collars, pyjamas, overalls, etc.

Eighty-five per cent of the rag trade was now in Jewish hands, and one in three of all workers living in the Rothchilds Buildings in the East End in 1900 were classified as employed in the garment industry. The trade was growing from strength to strength. The major divisions came between bespoke tailoring and ready-to-wear, between ladies' and gentlemen's clothing, and between men's work and women's work. As well as a garment industry in the East End, there was another in the West End, with the German tailors of Soho. It was mainly bespoke and high-class tailoring in the West End, as against mass production and sweat-shops in the East End. The tailoring industry was, like most contemporary trades in the East End, plagued by labour problems throughout the early 1900s, and with the arrival of the Jews into the rag trade, which increased the working population, the question of 'sweated labour' was raised.

Britain has traditionally been a source of refuge for immigrants seeking work and a decent life. Racial and cultural prejudice, language problems, low educational qualifications, and lack of skills made the immigrants easily exploitable. This was very apparent during the Jewish 'chain migration' of the late nineteenth and early twentieth centuries. The phrase 'chain migration' describes the tendency for immigrant people to go to areas where their own people are, for reasons of security, aid, and identification. This led to the mini-ghettos in the East End and the need to form organizations such as the Jewish Board of Guardians to help the newcomers.

From the thousands who swarmed to the East End emerged the 'homeworker' or 'sweater'. These were men, women, and children who worked morning, noon, and night in their own homes. In many cases, these people would toil for days without seeing daylight or breathing fresh air; and in the tailoring workshops, equipment and cutting tables left a minimal amount of space in which to work. In 1892 the average working day was between thirteen and fourteen hours. In Charles Booth's survey of *The Life and Labour of the People of London* (in 17 volumes published between 1889 and 1903) we are told of the earnings of operatives in the tailoring, boot, and furniture trades. Booth had the following three categories for the working-class population of London:

(1) The 'very poor', earning less than 18*s*. to 21*s*. per week.
(2) The 'poor' whose earnings averaged over the year between 18*s*. and 21*s*. per week.
(3) The 'poor' with earnings of 22*s*. to 30*s*. per week.

All of these were well below the poverty line, which was calculated at 30*s*. per week.

As early as 1874 the first tailors' organization was founded, but it lasted only a few weeks due to lack of support. In the years between 1892 and 1901 there were fifteen trade unions in operation amongst the clothing workers, although only nine were recorded as still in existence in 1901. The most substantial union was the United Ladies' Tailors and Mantle Makers, established in 1891. It had 550 members and kept a regular membership up until the First World War. The Independent Tailors, Machinists and Pressers Union and the International Journeyman Tailors, Machinists and Pressers were amalgamated in 1898 and had several hundred members.

What was perhaps the first strike led by Jews took place in Leeds in 1885, when garment makers met in a local synagogue to demand a one-hour reduction in their thirteen- to fourteen-hour day. They struck for a week and won their case. Jewish trade unionism reached its peak in 1890, when 10,000 London tailors rebelled. This led to a general strike of tailors and 'sweaters' led by an English Jew, Lewis Lyons. Their demands included huge reductions in their hourly day, an introduction of lunch and tea breaks, and controlled wage rates. There was also a northern Jewish trade union, organized from Leeds, but it appears to have had a far more practical base, its prime concern being the maintenance of an acceptable standard of living. It also anglicized its image, whereas the London Jewish unionists segregated themselves and refused to assimilate with their gentile contemporaries.

The women's mantle business which had been introduced to the East End by the immigrants had by now developed into a major industry and included dresses and other ladies' wear. Clothing firms like John Collier, Alexandre, Weaver to Wearer, Harella, Gor-ray, Simpsons, and Cojana were all built up by the Jewish immigrant pioneers. Alexon was founded just before the First World War by Jewish business men Sidney Drake and Reggie Adler. By manufacturing reasonably priced, well-made clothes at a time when the increasing emancipation of women had strengthened their spending power, the firm was firmly established by the late 1930s. During the Second World War it produced uniforms and then grew from strength to strength. A 'trouser for ladies', called Slimma, was added to the Alexon range and was at first worn by only the most fashionable and liberated women. When slacks, as they were called, became acceptable in the 1950s, Slimma's success was secured, and the Slimma Group clothing operation was established. Subsequently a merger with Tootal took place, establishing a major ready-to-wear clothing concern. Turnover of around £70 million was reached in 1955, its first year.

By the early 1950s many rag-trade companies had moved from the East End to the West End, within the boundaries of Oxford Street, Great Portland Street, Euston Road, and Tottenham Court Road. Among the first to move was Cyril Kern, the managing director of the Morris Skirt Company, later renamed Nadler. Two more entrepreneurs were Monty Passes and Monty Burkeman of the Helene of London Group who moved to showrooms off Regent Street. They took under their wing a young hairdresser from the East End whom they thought had 'chutzpah' and talent and set him up in a small salon off Bond Street. His name was Vidal Sassoon. Rensor, Polly Peck, and Norman Hildebrand are but a few of the numerous businesses that moved west, having once started off in the ghetto workshops of East London. Their grandparents would probably never have imagined that their successors would become multi-million-pound corporations.

These great Jewish businesses established in the nineteenth and twentieth centuries are fast losing their distinguishing Jewish characteristics. An oil painting of the patriarch may still preside over the boardroom table, or in the case of Rothschild and Marks & Spencer there are still members of the family to keep the traditions alive, but in the main the classic figure of the Jewish entrepreneur has been lost. At Burton a management consultant is in control, and at Tesco (originally a Jewish firm that made and sold clothes) the US financial consultants McKinsey are in control. Russo-Jewish immigrants were responsible for the massive retailing multiples and chains: Marks & Spencer, Burton, Great Universal Stores.

One finds oneself asking why was it the Jews, and why did they have this effect on the country? Though the Jews are not by nature better at business than others, they had an immense drive and built-in determination to succeed. They had inherited a whole system of values which, when combined with their circumstances, led them to operate competitively.

Today we find ourselves living in an increasingly multiracial society, and it is commonplace to find Jews, Afro-Caribbeans, Indo-Pakistanis, and Cypriots living in mixed communities rather than in segregated mini-ghettos. The Jewish heritage in the East End has all but disappeared. Despite the vibrant quality of Jewish life in the East End, the immigrant quarter became a frustrating environment, with its overcrowding, lack of open space, noise, and congestion. The increasing prosperity of the Jewish community led its members to move to the more salubrious suburbs of Stamford Hill, Golders Green and Edgware. The bombing suffered during the Second World War acted as a catalyst in the East End's decline; and Stepney's general population fell by over half, from 200,000 to 90,000. The pace of Jewish dispersal was hastened by the expansion of work opportunities in areas such as Ilford and Romford compared with the decline of employment in the East End.

In the East End today practically all of the once Jewish shops, houses, and meeting places are now owned by Bengalis. In 1960 the Bengali population in the area around Brick Lane and Whitechapel was estimated at 3 per cent; by 1980 it had risen to 90 per cent. Like the Jews, many Bengalis dispersed to northern and provincial areas – for example, Bradford, Leicester, and Birmingham – but the majority came to East London in the familiar 'chain migration' that had occurred seventy years previously. Many of the racist charges now made against Indo-Pakistanis echo those formerly made against Jews.

The Indo-Pakistanis came at a stage when the East End was economically depressed, partly as a result of industry moving out and partly due to the dispersal of the Jewish community. They thus had an opportunity to improve the area and establish their own businesses. It is interesting to see how the clothing industry has changed. The Jewish 'outdoor' workers or homeworkers are a dying breed and have been replaced by hundreds of Indo-Pakistani workers in East London and by Greek Cypriots in North London.

It is not difficult to see why the Jews were prepared to struggle so hard to become their own masters. Most could see the attraction of owning their own small business rather than working for someone else. What did they have to lose? They were so low on the economic scale that they could afford to take risks in the clothing trade and in business. Their lives had been fraught with disaster and tragedy, and they learned to seize chances when they could. Because they had to suffer persecution for so long, they built their own self-defence system and a desire for independence and freedom. Perhaps this same motivation provides the Bengalis with a similar determination to make the East End and their rag trade a success.

SECTION II
Fabrics and Construction

INTRODUCTION

Fabric is an intrinsic element in any study of the fashion industry, since textile manufacture acts as a service and is often a crucial factor in the emergence of new fashions. Fabrics have a history of invention and innovation of their own. But when seen in conjunction with their use, broader issues emerge; and the study of these issues contextualizes both the fabric and the clothes constructed from it.

The three sources which are utilized to make fabric are plant, animal, and chemical, and the different ways of manipulating these resources create varied results. The term 'fabric' encompasses all materials, whereas the term 'textile', which is defined as a woven fabric, excludes materials such as leather, which is a skin, and latex, which can be formed as a film. However, the majority of fabrics are produced by a weaving process which consists of fibres meshed together to create a cloth. The different types of fibre and the different methods of production make for varying results, and the range is increased by the use of colour, pattern, and texture. This section of the book illustrates the many possible treatments that the subject of a particular fabric can inspire.

The first essay (Chapter 5) attempts a methodology of approach which is more commonly appropriated by fine art rather than an applied art, in that surface decoration is analysed. Representation in pattern is a form of decoration with which fashion designers have a lot of contact, since they often design the print but not necessarily the basic fabric to which it is applied. In the case of woven fabrics, lengths of cloth are manufactured, from which a garment is cut. Using the process of knitting, an alternative method of construction becomes possible; the garment can be fully fashioned whilst it is being made. The unprecedented decline of the British woven textile industry in the post-Second-World-War period is not just an indication of social

Figure 3 Advertisement for 'The Scotch House' (1987). In the 1980s traditional indiginous fabrics such as Harris Tweed are enjoying a revival. This advertisement illustrates how tradition can be redefined and modernized'. Nostalgia for a past era is conveyed by the use of natural yarns and dyes, and conventional clothing forms.

and economic changes in Britain. The decline is one feature of a global industry which was dominated in the eighteenth and nineteenth centuries by Britain but now depends to a greater extent on cheap labour in other parts of the world.

In the 1960s synthetic fibres seemed to be the economic solution to a flagging industry, but in the 1980s the emphasis is on natural yarns and fabrics using a proportion of chemical fibre only when it can be shown to be advantageous to the resulting garment. For example, a small proportion of nylon in cotton socks prolongs wear and elasticizes the fabric to enable a snug fit. Simultaneously with the current interest in knitted fabrics there is an upsurge of interest in traditional woven British fabrics such as Harris tweed and Tartan.

A knowledge of textiles is essential for a fashion designer, because the designing process and the finished product rely heavily on the fabric choice. The essays in this section therefore illustrate how academic work can be directly relevant to design practice and can offer a more sophisticated approach to design problems.

5

An interpretation of visual images in relation to textile design

ANDREA PAINE

This essay deals with visual images in relation to textiles. The type of textiles discussed are those connected to commercial design, with more emphasis on fashion than on interior design, thus relating textiles to the human figure. This aspect has been chosen for consideration from a fashion designer's perspective, rather than from that of a textile designer. A textile designer is generally more fine-art orientated, and his or her approach would be in terms more of display and expression than of wearability and commercial viability.

Various visual images can be used as inspirational sources for designing textiles. These images may be of anything that appears to have potential for a design. The object-matter chosen may range from natural surroundings to forms of art, books, television, photography, and man-made structures – practically anything that has been seen that may give an idea. To be able to use these sources for a design, they are generally captured by a two-dimensional form such as a sketch, painting, or photograph. Magazines and books often have illustrations which can be used for textile design purposes in the form of either repeat shapes, colour combinations, or compositions. In other cases the possibilities for design may have already been used by artists but not by textile designers. Artists' interpretations may be in expressive forms such as paintings, prints, postcards, and photographs, but the textile designer uses these images and involves them in his or her own work.

In the context of art and design, textiles are neither paintings nor sculptures, but somewhere in between, involving both elements. So, when designing, many considerations are taken into account: colour, texture of fabric, pattern, repeat, method of execution, and also how the fabric is cut after having been designed. For the designer, there is the problem of how to translate ideas through the right medium and technique to produce the best design from the subject chosen. With the use of modern technology, the effects produced on fabric are different, and the manner of execution has also changed; but in the end, whether textile designs are produced by means of hand, wood-block, screen, roller, or computer, a similar thought process has still been in operation – from idea to finished work.

The textiles produced are generally for either of two main uses, interior décor or clothes design. Interior design gives emphasis to the shape of the room and furniture, and fabric designs are intended to complement the character of the room. The basic requirements are functional. When designing fabric used for dress, however, the emphasis is much stronger on the design itself. It is still used around a three-dimensional form but this time it is a moveable form – the human body. The cut of the fabric, drape, and folds are all contributory factors in relation to movement. The scale for dress fabric generally has been towards smaller proportions, since the surface area is less than that of furnishings. But recently designers have been mixing these uses; furnishing fabrics have at times proved popular in dress design, indicating that furnishing fabrics themselves have become more adaptable and that dress designers have looked to larger-scale designs for their textiles.

When fabric designs – dress or interior – are compared to the look of natural patterned objects (such as pebbles) the lines and shapes of the designs themselves describe the whole form of the object, area, or person rather than merely the outline. This effect is often used in camouflaging the real size of a large person (with the use of optical illusion, where the eye is drawn downwards with a succession of vertical lines, for example). The effect is to draw attention to the pattern first rather than to the shape. Besides the consistency of the texture of the fabric, the colour and pattern arrangement are the most noticeable. The pattern is not the only factor displaying any visual information but is only part of the whole look of the object. The tie in an outfit is seen as part of the whole, co-ordinating with the rest. But even though it may sometimes play a very small role, the rhythm and overall concept are still the major consideration for the viewer.

Designing fabrics works as a unit-building process, making a large construction from many small elements. The fibre content itself, as well as the step-by-step process of producing a pattern, contributes to the design. The idea of a repeated shape may be one of the first ideas that comes to mind when designing textiles, but texture is equally an important consideration. Fine-art painters such as Seurat and Jackson Pollock have created all-over textures within which there is form; shapes emerge from the form rather than being deliberately created.

The effect created by a pattern printed on one type of fabric may look totally different when printed on another. Different fabric weights and textures give totally different looks. The fabric acts in a similar way to an artist's canvas, but instead of being flat it works as a skin – around a sculptured form. On this moving surface the yarn often appears different with changes of natural and artificial light. These effects are often enhanced by means of synthetic yarns with a shiny or metallic appearance. The arrangement of fibres produces a design within the fabric, rather than printed on the surface. This woven method enables designs to be integrated with the fabric. The change of colour appears only in the yarn – that is, along the length or width of the fabric. A very intricate patterning can be achieved in stripes and checks. These, being more traditional designs, can be done almost mathematically. The designs, therefore, work on colour, fabric content, and composition rather than on shape or pattern association.

Whether woven, printed, or appliquéd, the design considerations for textiles can produce a unique combination. Few other forms of design are so versatile in terms of practicality, movement, texture, composition, shape, method of production, and colour. But however the fabric has been created, the colour is the most significant element. Different placements of colour within a pattern can drastically change the whole mood. Colour, more than the pattern, texture, or overall design, has the greatest attraction when viewing a fabric.

During the late nineteenth and twentieth centuries there has been a contention between designers' ideas and their influences. The debates began with William Morris in the 1870s when he rebelled against the effects of mechanization in design. It was on this premiss that the Arts and Crafts movement was formed. Later this progression led to the more commercial Art Nouveau period, and artists such as Charles Rennie Mackintosh produced floral-inspired textile designs at the turn of the century. Italian Futurism opposed these ideas and emphasized instead new forms of beauty inherent in the machine age – fast cars, trains, and components of industrial production. Russian Constructivists worked on similar ideas but with an eye to creating architectural, textile, and ceramic designs for a new, communist society. New principles were set for a type of art and design in which politics was seen to be one of the motivations.

The Art Nouveau, Futurist, and Constructivist movements show how art and design ideas have derived from pastoral, technological, and political ideas. Organic and natural influences can be seen in the work of artists such as Morris, Mackintosh, and Raoul Dufy, all of whom used plant and floral forms as their inspiration, producing work on a decorative basis using harmonious flowing shapes. There have been many revivals of this naturalistic form of expression, in the 1960s as well as amongst current 1980s designers. In sharp contrast to these naturalistic patterns, Russian artists of the 1920s were designing textiles to echo the revolutionary transformation involving technology and mechanization. The images used were often from industry, not merely for decorative but also for propaganda purposes, emphasizing the importance of the revolution. Political images have continued to the present day in fashion in the work of designers such as Katherine Hamnett, whose T-shirts frequently bear slogans.

Textile designs for clothing are often based on check, stripe, or spots in various forms, such as dog-tooth check, tartan, herringbone, pin-stripe, and polka dot, all of which appear again and again each year in various forms according to fashion trends. Often these traditional influences are popular as fabric designs for clothes; but other influences are becoming fashionable which have traditionally not been related to textiles. These may be from areas such as interior design, architecture, and fine art, where new fashion trends are often seen before they are transferred on to fabric. One of the latest styles of architecture is 'post-modern classicism'. This has evolved only in the last decade or so, combining classical shapes and forms with modern methods and styles. An example is the Piazza D'Italia by Charles Moore in New Orleans. Interpretations of these classical ideas into textile form appeared on fabric for winter 1985–6. The Paris exhibition 'Première Vision' in October 1984 featured these ideas, with classically influenced imagery of vases, columns, and statues.

Textile producers today use a variety of sources of design. They choose work of different artists in accordance with their own styles and fashion predictions. Companies such as Liberty, Habitat, and Laura Ashley produce textiles for clothes and interior design stamped with the company's own identity. The variety of uses textiles can be put to mean that when co-ordinated they can produce an overall design image. This parallels the way artists such as William Morris and Sonia Delauney made curtain fabrics, wallpaper, cushions, and in Sonia Delaunay's case coats, all of the same textile design. The development of a style in a company or design group sometimes extends to areas such as furnishing, graphics, crockery, bed linen, and perfume in order to create a complete life-style and visual appeal. Liberty's is a store of long standing, which has always produced a multitude of traditional and modern textiles. Traditional Liberty prints can be bought by the metre or made into garments, sofa covers, and toiletries. The same fabric print can be seen on kitchen items such as trays and cooking equipment. Habitat, with emphasis on housewares, has similar ideas. Ranges of furnishing fabric designs are both printed on fabric, as well as on paper for stationery, and also used for window blinds. These all go to make a co-ordinated living environment. Laura Ashley works on similar principles, but the prints are mixed together to create a 'look' in clothes and interiors. The 'classic Laura Ashley look' proved so successful that it enabled the company to 'go public' in autumn 1985.

Recently there has been an upsurge of new imagery in printed textile design, becoming as popular as knitwear design. Until a few years ago inventive textile design appeared to be too risky to be mass produced. The manufacturers would rather not 'stick their necks out' for the fear of producing an unpopular print design. This may still be their policy, but there are now enough young designers willing to experiment and create their own new individual prints. How long this will last is difficult to answer.

These new approaches depend on the mechanical processes by which fabric is manufactured and printed. Wood-block printing has been largely superseded by silk screen, also practised by hand but enabling a large area to be printed at one time. This works by a stencilling process, whereby printing colours are pressed through a fabric mesh covering a frame, and areas not to be printed are blocked out. Paper stencilling would be the simplest form, but a photographic method has enabled designs to be blocked out by using chemicals to fill the correct holes in the screen. This method is still in operation today with small companies and individuals requiring high-quality and short-length printed designs. Hand printing is the best method, but it is time consuming and uneconomical. There are variations of the screen-printing process, with semi- and mechanically separated screens, the newest using computer-aided design (CAD), but both imitating hand-operated printing.

Apart from changes in printing technique, chemical developments have made possible the production of synthetic fabric. Synthetic fabrics are used extensively for interior design since they tend to be more durable. For fashion uses synthetic fabric has gained some popularity due to the low prices and easy care of the material, but its 'feel' is often inferior to that of natural cloths. Synthetic fabrics first became popular in the 1960s and have recently returned to fashion partly as a result of improvements in fabric constructions and combinations with natural fibres. This has led to many designers, for example Body Map, taking advantage of the synthetic fibres' properties. Major companies such as Courtaulds have in the past printed only large quantities for department stores like Marks & Spencer, since smaller amounts would be uneconomic. Recently, though, Courtaulds has been printing small amounts for younger textile designers, since such experimentation with fabrics pays off in terms of exports abroad. Courtaulds' fabric is unable to take a pigment – that is, a colour printed on top – so dyeing procedures must be developed where colours are absorbed into the yarn itself. This process is achieved with much greater difficulty than it is on natural fibres, because the synthetic yarn is non-absorbent.

Apart from the fabrics themselves, the visual imagery used from computers as a design source for textiles is only the start of a different type of textile design. Up until recently it was possible to design only using a grid system, but the designer can now draw virtually straight into the computer. He or she draws a line or mark by hand on a level board where the image is recorded and passed into the machine. This appears on the screen and is recorded. The computer can then be programmed to repeat the image, reverse it, and rearrange scale, or whatever the designer requires. Hopefully art schools and fashion departments will soon be able to afford CAD textile machinery so that textile and fashion-design students can experiment with textile imagery before entering industry or setting up on their own.

6

Knitting: the fabric of society

JULIE ABBOTT

The British Isles have a long and varied tradition of knitting. This tradition has many facets and can be divided into a number of different study areas: domestic hand or machine knitting, industrial hand or machine knitting, and regional or designer knitting. This essay concentrates on the social and cultural implications of domestic hand knitting.

A study of domestic knitting shows the practice to be as diverse as the society in which it exists. At one end of the social spectrum in the nineteenth century there were the genteel ladies of the upper classes and the *nouveaux riches* who filled their leisure time pursuing the domestic arts and crafts. At the other end of the scale were the lower classes, who regarded knitting, as they always had, as a means of creating a garment which was both practical and economical; these attitudes were typical of small communities whose work demanded strong and durable outer wear. Communities like the fishing villages of Scotland created their own particular form of knitted sweaters.

With the rapid growth of industrial society in the nineteenth century, and the migration of workers from the home to the factory, there followed a decline in the 'leisure' time of the working classes and therefore a reduction in home activities like handicrafts. A law had been passed in 1872 enforcing the teaching of hand knitting in schools, but by the advent of the twentieth century the majority of knitting was done by the older members of the family. Working-class knitting changed little from the beginning of the century to the 1920s and was still largely undertaken out of economic necessity.

At the end of the nineteenth and beginning of the twentieth centuries knitwear was an obvious choice for the expanding ready-to-wear industry in so far as before its advent women could not possess a knitted garment without having to spend their valuable time making it. This led to the growth in popularity of manufactured knitted garments on the one hand, and a further decline in domestic knitting on the other. The fashion designer Chanel, who had enormous success with her knitted jersey collections in the 1920s and 1930s, was a major force in developing fashion design in this direction.

Traditional knitted patterns became fashionable in menswear in the 1920s. The Prince of Wales, a leading fashion-setter of the time, popularized Fairisles worn with plus-fours, and Argyle knitted socks. This vogue for men made for some of the most colourful and decorative male fashions in Britain this century.

With the Wall Street Crash in the United States in 1929 and the subsequent world economic depression, the 1930s became a fruitful time for home knitting due to poverty brought about by unemployment and the need to make clothes rather than buy them. Developing simultaneously with hand knitting in the 1930s was an increasing wealth of instructive literature. The period was dominated by women's magazines such as *Women and Home, The Needle Woman,* and *Home and Fashions*, which assumed the role of previous knitting books by publishing patterns for all types of knitted items as well as hints and tips on their maintenance. Knitting patterns with less complicated instructions were a great help to the home knitter, published by an ever increasing number of spinners such as Patons, Baldwins, and Sirdar. In 1931 *Vogue* published knitting patterns in colour for the first time and in 1932 introduced a series of *Vogue* knitting books which survived until the 1970s and were renowned for their high fashion content and for the clarity of the instructions.

With the onset of the Second World War the 'Great Age of Knitting' came to an end as a result of women being encouraged to enter factories and assist the war effort. Britain now moved into a period of utility and rationing. Knitting for pleasure took a back seat to knitting 'comforts' for the troops; and since virtually everyone had a relative or friend in action, the civilian 'army' of those 'knitting for victory' soon grew to epidemic proportions. These activities were encouraged by many charitable organizations, such as the St John Ambulance Brigade and the Women's

Royal Voluntary Service, which donated wool free, with the completed garments to be collected for dispatch from depots throughout the country. All manner of groups and even schools held 'knit-ins', often adopting a particular ship or platoon to which to send the completed pieces. Knitting patterns were printed on plain leaflets similar to those used today, with most issued by the 'Comforts Committee', a government body which co-ordinated knitting activities to the best advantage of the war effort. Rationing meant that almost everything was purchased on receipt of coupons, which were regulated and issued by the government to avoid the threat of shortages. Although wool needed more coupons than rayon or even silk, it remained the most popular yarn throughout the war years, due largely to its ability to retain body warmth, which was essential in a country of widespread fuel economy. In such a state of emergency, steel became precious for use in munitions and arms factories, so wooden needles replaced steel for the duration.

Knitted garments remained popular among the civilian population also, with short-sleeved styles being deemed fashionable and conforming to utility regulations. In addition to its warming qualities, knitting remained popular because old sweaters and cardigans could be unravelled and re-knitted into more up-to-date styles. By clever use of colour and surface decoration, knitted garments were often a refreshing change from the more homogeneous utility clothing. A prime government advocator of this resourcefulness was a publicity lady called Mrs Sew-and-Sew, who, under the slogan of 'Make do and mend', issued knitting patterns and leaflets with tips and ideas for the housewife, such as unravelling two dishcloths to make a jersey and absorbing odd bits of left-over yarn into Fairisle knits. Wool spinners and magazines published patterns in standard-weight yarn to allow the knitter more flexibility, and economical ways of using knits became popular, like knitted gloves, hair snoods, brooches, and scarves. In 1942 the Board of Trade published a statement warning of the imminent shortage of woollen stockings for men. New styles of short socks became fashionable and took their place. Also in 1942 a popular style was the Victory jumper, which was made with everything in a V-shape: V-neck, V motifs, and V patterning. This last vogue is an indication of how knitting was as much a morale booster for those at home as it was for the troops. Many people who had never knitted before did so to occupy themselves during air raids when the black-out meant that reading and other hobbies were impossible.

The end of the hostilities in 1945 left Britain devastated by the effects of war, and people's leisure was mostly taken up with rebuilding and readjusting to peacetime. Rationing continued until the early 1950s, with the age of austerity coming to an end (in theory at least) only with the Festival of Britain in 1951. A popular style which originated at this time was the chunky sweater or 'Sloppy Joe'. The Italians were the first to incorporate this sweater into fashion, but it was popularized by Americans who combined it with casual jeans and sneakers to create a look which has remained ever since. The large sweater became fashionable after the restriction of size imposed during rationing, just as after the First World War it won acclaim as an androgynous item of clothing which was both comfortable and practical to wear. With the new thicker yarns and the popularity of luxury yarns like mohair and angora, a 'Sloppy Joe' could be knitted remarkably quickly and jazzed up with jewellery. The sweater has remained consistently popular ever since.

The 1950s were a time of continuing the advancements which had been made in textile science and technology before the war. In 1950 the British Wool Marketing Board was set up to organize and regulate the industry. After the introduction of nylon in 1939 – the first completely synthetic fibre – successive developments were delayed during wartime. By the 1950s, however, renewed interest was being shown by both public and manufacturers in experiments in fibre technology. Many new fibres, such as Orlon, were developed, which were originally intended as a substitute for wool and have come to be combined with natural fibres to exploit the good qualities of both. These developments meant that the home knitter now had a much wider choice of yarns, many synthetics being made available at cheaper prices than wool. The mixes meant that the knitter could now have all the good qualities of the natural fibre, such as warmth and durability in wool, but with none of the problems of 100 per cent wool, such as shrinkage and felting. These achievements revolutionized the old manufacturing techniques, and the combination of the synthetics with existing and new dyes led to entirely new hues and intensity of colour being produced.

Along with this new technology came improvements in machinery associated with knitting. The 1950s were the period when the realm of the home knitter was invaded by the knitting machine. Home knitters were as sceptical about using it as people had been about the sewing-machine a hundred years earlier. It was primarily the formation of Knitmaster in 1952 and the introduction of its 2500 model that persuaded the public to take notice of this new phenomenon. Subsequent improvements over the next few years meant that hand knitting was gradually replaced by the machine in many homes.

Although many people still saw hand knitting as a form of relaxation which happened to produce a garment at the end, during the 1960s there was an escalation of machine knits as hand knitting fell to an all-time low.

The first indications of a revival of interest in the aesthetic values of knitting occurred during the 1970s, when a handful of designers included a large proportion of knitted garments in their collections. However, the majority of these designers dealt with machine knitting, and although they did much to enhance Britain's international reputation for superior knitting designs, the first signs of hand knitting being successfully adopted by *haute couture* were in 1981, when Ralph Lauren produced the first of his Prairie, Navajo, and Pioneer collections using knitwear derived from Navajo or traditional American samplers. Lauren redefined the popular image of knitting and helped turn the 1970s ethnic look into 1980s chic.

Elina and Lena are a British designing duo who continued in this vein by basing many of their knits on traditional Fairisle and Aran pattrns. The subtle yet classic elegance which their knits represent has encouraged a world-wide interest in contemporary British knitting. There are many other talented British knitting designers who have also contributed to the 1980s knitting revolution, and specialist designer knitwear shops have become established as a result.

It is ironic that, given its humble origins and struggle for popular recognition, hand knitting should have evolved to a point where it and the *couture* ready-to-wear industry have become essentially dependent on each other. As a result the over-pricing of designer hand knits makes them inaccessible to the average consumer, and knitwear has become an item of luxury for many of the working classes. However, most of the successful designers actually rely on working-class 'outworkers' to knit up their designs. Indeed, many designers use their outworkers as a selling strategy; for example, Rococo stitch a label into every garment stating 'This woolly has been knitted by the fireside of . . . '

The relatively sudden output of talented British knitters stems largely from the reintroduction of the teaching of creative knitting in art schools in the 1970s. Many of our most respected designers in this field are graduates from such courses, where they were taught the technical side of knitting as well as design.

In the 1980s there is undeniably an increasing awareness of fashion on the part of the familiar home knitter through the availability of magazines, books, and television programmes. The gap between the two areas has been narrowed for the most part by a small batch of designers who successfully anticipated that the interest in designer knits would cause people to return to hand knitting themselves. Their actions have made *haute couture* potentially accessible to anyone who can knit, by providing the designs and the inspiration.

7
Tartan

SALLY STACY

Tartan has appeared throughout fashion history, whether worn conventionally or (more recently) in new and different ways. During the twentieth century tartan has been revived a number of times, sometimes with only a few years' gap in between each appearance. Its resurgence has taken many forms, because tartan is versatile. This versatility is demonstrated through tartan being used not only by *couture* fashion houses but also by Punks and other more marginal groups on London streets.

Certain fabrics relentlessly return and enjoy a revival, often revitalized by designers according to fashion trends. For example, when Kenzo had the audacity to print bright, clashing flowers over the top of tartan, he did it with complete disregard for the tartan, and yet as a fabric it was ingenious.

Innovation in design occurs in combining the different characteristics of particular materials. This is where skilful designs and inspired designers are at their best. It is then that a strong fashion statement is made which everyone wants to wear, and yet retain some sort of individuality.

The old idea of creating a completely new look in fashion has become less and less what designers strive for in their work. Fashion now relies much more on picking over clichéd garments or fabrics and changing them, either by juxtaposing them with something unusual or putting two opposites together and creating something new. The ability to re-create a look or image from something old is a particular talent and one of the main components of fashion today.

Fashion in most cases has to have some sort of foundation from which to work and to build up ideas, whether it is a textile design or a garment design. Tartan is a good example of a fabric which has certain regular features, almost rules, but which can be adapted and used over and over again in many different forms – as can Paisley, stripes, dots, and herring-bone patterns.

Traditional and anonymous designs are the two major qualities that a material appears to need in order to be reused as a fashionable textile. Tartan's history denoting the stratification of clans in Scotland and its subsequent use as a military uniform mean that for some it retains an image of dignity, correctness, and order. It may be because of this that the custom arose for tartan to be used in school uniforms and children's clothes in general.

Tartan's constant 'reappearance' has made it something of a cliché within high fashion. This is not necessarily a bad thing, since its clichéd, jokey elements are often selected by designers in a light-hearted manner in order to add to overstated, flamboyant garments.

Tartan is not just a fabric with a particular geometric pattern. There are other issues which are relevant to its social and economic history, together with its particular fashion and textile history. This can be seen most readily in the adaptation of tartan as part of Punk fashion. Since this checkered fabric has a patriarchal image, it became one of the vehicles by which the Punk movement could express its discontent with and reaction against society. By mistreating this previously sacred fabric, by wearing it in very unorthodox ways, and by unconventionally combining it with other, unexpected garments, the fabric itself became transformed, and thus its meaning changed.

8

Leather: necessity or luxury? From workwear to couture

ISOBEL LEADLAY

Leather was originally used instead of cloth to make certain garments because of its strength. The first aprons were made of a whole animal skin, using one folded corner of the skin as a bib. For shoemakers and cobblers, leather aprons protected them from the rubbing of the black cobbler's wax they used. Tanners and slaughtermen both used aprons as a shield against splashing, and also against knife injuries. Metalsmiths and coopers wore aprons to save them from sparks and friction by the metal.

Throughout the eighteenth and nineteenth centuries, vegetable tanning was used in processing leather; it was not until the late nineteenth century that chrome tanning came into use, shortening the tanning process of leather from weeks to days. Leather slowly moved into other fields of dress besides simple working aprons – for example, jackets and coats. It has taken over half a century for people to realize how versatile leather is and to incorporate it into everyday casualwear fashion.

Leather and suede have been much used in fashion since the 1950s, but apart from jackets and coats their popularity has been limited. It was not until the late 1970s that Italian designers noticed their further qualities, and they promoted the use of supple leathers and suedes in a vast range of colours. Some of the skins were printed on, others embroidered and used for women's blouses and dresses, as well as the casual, more obvious style of clothing which was usually associated with the tough, hard-wearing outdoor coat, jacket, and trousers.

Although it was expensive, many women found leather appealing, and a permanent section of the fashion market became established for suedes and leather, hence leather in *haute couture* fashion. Of course, the style, shape, and colour of these outfits change every season, but there is usually a more everyday, practical style which changes only slightly each season, so that the costume does not become dated so quickly, especially since it is so expensive. *Haute couture* designers from all over the world are now designing more and more in leather.

Leather is a versatile fabric which can be and has been used in many different ways. Whilst having a place in the world of *haute couture*, it also has a place in the history of 'anti-fashion'. For years, leather and, more recently, rubber and PVC have been associated with fetishism or perversion, in partnership with bondage, whips, studs, and a more sado-masochist dominant-master-and-slave appearance. The characteristics of leather which lend themselves to the 'seamier' side of fashion are not merely to do with its exclusivity. It is hard to explain what it is like to wear a leather or rubber suit, trousers, or skirt, for each person experiences its body-hugging qualities differently. A fetishist would be able to describe the sensation of wearing leather more convincingly than someone trained in 'mainstream' fashion, for whom the feeling of fabrics is to do with both their sensation and the external appearance and cut.

Putting on a dead skin, easing the body into a leather suit, is a different experience for one person compared to another. For a leather fetishist, the smell, feel, and touch are erotic stimuli. It is real, supple, and smooth, just like our own skin, except that while it conceals and hints at the revelation of flesh it doesn't rely on the exposure of the possibly flawed human body. The fact that it was originally part of an animal becomes, for the fetishist, an added pleasure – that of being inside another creature's skin – and apparently gives the wearer a feeling of power, like a hunter.

Undoubtedly, leather's biggest association is with machismo. It is tough and durable. Soldiers and warriors throughout most nations' histories have worn leather in combat, quite often with accompanying studs and accoutrements for both functional and decorative purposes. Leather jerkins dominated the look of the medieval European man. The blacksmiths and labourers who have always worn leather aprons have predominantly been male. It was not until a better process of tanning and understanding had been found that leather could be made lighter and more flexible – in effect, considered more 'feminine'.

Figure 4 'Gothic' punk (Photo: Marek Walisiewicz). This is one example of the fragmentation of punk which occurred post-1978. The leather 'bikers'' jacket and the use of the colour black have remained potent iconoclastic symbols within many subcultures.

Leather can be dyed any colour and worn thick and warm for the winter, or thin, light, and 'slinky' for the summer. Shoes, hats, jackets, swim-suits, shirts, even socks – anything is possible for leather has proved to be one of the most versatile of materials, if not *the* most versatile. There was a time when to describe leather as 'feminine' would have been a contradiction in terms, or to suggest a leather jacket as ideal for the summer would have been ridiculous. But now all that has changed, just as the 'rules' of leather are changing still.

Leather is for anyone, for any sex, for any social stratum. It has gone from being working class, through *haute couture* at the higher end of the market, to being antisocial, then emerging in mainstream fashion. And it has benefited stylistically along the way.

SECTION III

Marketing Strategies

INTRODUCTION

The point of sale is the end of a process whereby fashion becomes consumer orientated. Since the Second World War, marketing strategy has taken an increasingly important part in reducing the high-risk nature of the fashion business. Certain types of retailing (such as department stores and boutiques) have radically altered patterns of trading in the twentieth century. The nature of the market-place is changing from one based on small high-street shops to a more complex variety of outlets. In particular, chain stores have increased in scale in recent years, often putting small shops out of business. But these changes do not occur in a vacuum. Consumption does not increase merely because a new branch of Next or Marks & Spencer opens on our high street. Increased consumption of material goods, such as clothes, relies on a social system which encourages us to believe that more and more possessions are essential for our 'good', and that to attend two social occasions consecutively in the same garment would be a signal of our incapacity to be original or innovative with our appearance. In this way, appearance and 'creativity' begin to take on new meanings. The marketing of fashion is concerned less with 'new' products than with stylized presentations of existing fashions which suit current trends. Nostalgia for a previous era and/or inspiration based on cultures not our own, for example Aztec, are constantly reworked in fashion.

Marketing is a complex process from which many different forms of retail practice emerge. Business studies have increasingly become recognized as an essential part of all design courses, and yet more often than not they are unrelated to the cultural contextualization of the *meaning* of retail trade, promotion, display, and packaging. The history of particular department stores and the economics and politics surrounding the use of the

Figure 5 Window display at Top Shop: the 'Pepsi and Shirlie' collection, Oxford Street, London, 1987 (Photo: Lee Wright). The high street has become a unit of standardization. From 'Top Shop' to 'Next', certain fashion outlets predominate. Concepts and themes are the most common form of window dressing. In this example pop music and fashion have been co-ordinated to reinforce the importance of the public image.

boutique in the 1960s, for example, are as important in the teaching of business studies to design students as are items such as how to complete tax forms in the event of becoming self-employed.

The first essay in this section (Chapter 9) looks in detail, from both a designer and a consumer point of view, at the management and design policies of Marks & Spencer in comparison to Next, as examples of two successful high-street clothing stores. The second essay (Chapter 10) examines shop-window display. Display of the product is crucial to the business of selling, and fashion has always had a variety of outlets for presenting clothes, whether it be on the catwalk as symbolic performance presentation, or statically on the pages of magazines. But retailing depends significantly on shop-window displays, which represent an interesting combination of performance and static presentation. The third essay (Chapter 11) questions whether the use of particular sources for designing is a market

Figure 6 Advertisement for 'Benetton' (1987) (Photo: Oliviero Toscani). Fashion uses a wider and more diverse variety of references than any other design discipline. It selects from the gamut of cultural information and translates it into a form appropriate to its own structure. In this sense it is marketing global culture.

strategy to create interest in a flagging industry which has exhausted any new ideas.

In the last five years business studies have become an integral part of many fashion-design degree courses. As courses and students have adjusted to this discipline there has been an increase in the number of final essays which reflect the nature of their studies. This section illustrates that change, which the study of design history and theory needs to accommodate, because it is an aspect of popular culture dealing directly with the relationship between production and consumption.

9

How consumer demand is translated into production: Marks & Spencer and Next

LINDA WHITELY

The following extract was written in 1984. While it offers an insight into chain-store policy, it is not necessarily an accurate picture of Marks & Spencer today.

At the beginning of the nineteenth century, the pattern of shopping for clothes was much as it had been for the whole of the previous century. The goods were sold by individual shopkeepers, often craftsmen, making the goods they sold, or by private dressmakers. The Industrial Revolution affected the fashion trade significantly. The year 1815 marked the starting-point for a new age of production and a new pattern of retail distribution.

It was out of the nineteenth century that Marks & Spencer, probably the most successful chain store ever, emerged. A store which has grown, developing its shops, standards, and production techniques up to the present day, Marks & Spencer have an unusual and well-defined philosophy and body of principles, established by the founders of the company (Simon Marks and Israel Sieff) and by those who have followed them in leading the business. The most important of these principles are the attention they give to the public and the importance they attach to good human relations with their staff, their suppliers, their customers, and the communities in which they trade. Perhaps first and foremost, because they are in daily contact with customers and in frequent contact with their suppliers, Marks & Spencer are able to interpret to the manufacturer more or less what the customer wants, so the goods produced are those that are in demand. No goods, however high their quality, have real value unless they ultimately satisfy customer demand. At the same time, a retailer cannot succeed unless it has the right goods at the right time. Today, though Marks & Spencer do not own or control any manufacturing capacity, they work in close co-operation with some seven hundred independent suppliers which produce to their specifications. Nearly one hundred and fifty of these companies have been supplying Marks & Spencer for over twenty-five years. Fifty of these companies have been associated with them for over forty years.

Because of their close link with the textile manufacturers, Marks & Spencer have always been forerunners in the introduction of new fabrics, as in the use of the new synthetics which became available during the early 1960s. These fabrics had properties that no garments in the world had previously possessed; they were drip dry and crease resistant, had new bright colours, and were relatively cheap. By 1960 the girl who shopped at Marks & Spencer was said to be the best-dressed girl in town.

The special, if not easily definable, character of Marks & Spencer's stores and of the merchandise sold in them was already widely recognized by the public as early as the end of the 1950s. Millions of customers had a genuine sense that in buying at Marks & Spencer they were getting better value for their money than they could gain elsewhere. Largely because of this, a feeling emerged that in some way the company was performing a public service as well as conducting a business for profit. Today Marks & Spencer have 260 stores in the United Kingdom, totalling 634,000 square metres of selling space. Continuous expansion and modernization of the stores, along with improvements in merchandise, have resulted in the company's growth world-wide.

A design department was established in 1936. This was intended to keep the company abreast of the latest trends in fashion design, and also to create a team of experts and technicians in this field whose services would be available to their suppliers. Generally, chain stores do not have policies to set new trends in fashion. The majority of companies report that they do not attempt to create fashion, or to influence a change in consumer tastes from a fashion point of view – they are content to follow it; mass production of standard lines pays better dividends. A visit to the design room of Marks & Spencer shows that their stated policy concerning design is borne out. It emerged clearly from a conversation with the executive head of the skirts and lightweight suits department at the Baker Street head office in London that Marks & Spencer make a priority of knowing exactly where contemporary fashion design is going.

The company works very closely with the International Wool Secretariat and the Cotton Institute, which forecast colour and fabric trends in the form of colour charts and story-boards predicting textures and moods (usually five for every season). The designers and their assistants from every department travel around the world to the fashion capitals – Paris, Milan, New York, and Tokyo – visiting fashion and fabric fairs. Sample fabrics and garments are bought from designers, and overall 'looks' are gathered together. Visits are made to fabric fairs in West Germany and Switzerland, and from all these sources the fabrics are gathered and co-ordinated.

Department heads work closely together to create separates and suits which can generally be a 'mix-and-match' collection. The executive heads of each department make all the final decisions on colour, fabrics, quantity, and style. Each department has the same colour boards for each season, so they are guaranteed to produce complementary garments.

The sample garments are made in the design room and have to be approved by the executive head before any production possibilities are discussed. Production quantity is calculated in cost. The number to be made is decided from past sales. Most tailoring in the UK is very expensive, so the more complicated suits, jackets, etc., are sent to Israel and West Germany. Alexon and Reldan are two British companies that manufacture for Marks & Spencer, and they buy some top-selling items directly from Alexon and Reldan. These companies also have certain factories which produce entirely for Marks & Spencer. Marks & Spencer have a small team who circulate around the UK asking sales assistants what kind of problems the customers have recently been raising with them. Complaints about sizes, stock, and so on are reported back to head office, followed up, investigated, and more often than not solved – all this being part of Marks & Spencer's customer service.

It is still the custom for the counter space to be used as a laboratory for testing demand. If a new fabric is to be introduced, it is made up into an old, guaranteed selling style. The same happens to a doubtful new style; it is made in a guaranteed selling fabric. Sample garments are also given to staff at the head office to wear for three weeks for wearability testing. All the Marks & Spencer knitwear is designed from outside by designers who work in the factories. Menswear used to be designed this way, but it was felt that the garments had slid into a rut and had become too classic and rather mundane. New young menswear designers have been employed to work at Marks & Spencer's Baker Street offices to transform the firm's menswear and make it more fashionable and appealing to the younger man.

The Marks & Spencer technologists are also involved in the buying groups and not compartmentalized off. Today, though, there are no buyers in the accepted sense – they have staff called 'selectors' who are part of a buying team which includes specialists with access to centrally based expertise in technology, production engineering, and personnel. This is designed to ensure that, as far as possible, the different jobs are co-ordinated, and everyone is involved in quality control and upgrading. An important aspect of quality control is the attention given to detail. The company's garment production engineers are in regular contact with the clothing suppliers, alerting them to likely faults and recommending processes which will ensure that in general the finished garment is satisfactory.

To see this approach in practice it was necessary to visit one of the Reldan factories owned by Steinberg in Bishop Auckland. The quality controller from the lightweight suits and skirts department at Baker Street was going to check that the current range being produced was reaching Marks & Spencer standards and that the measurements were correct for each size – waist, hips, length, etc. This was tailoring on a mass-produced scale.

The process starts at the goods-in office. Here, every garment, quantity and quality of fabric, lining, and trim is recorded; the length of time it will take to produce the required amount is estimated; the deadline dates are set. A sample from each roll of fabric is taken from the middle of each roll, labelled, and sent to the Textile technology department. Each sample is passed through a series of rigorous tests after which it is repeatedly compared to a standard fabric sample. The pieces of fabric are then cut out and sent to the factory floor, fused with vilene by machine, and then ready to be sewn together by various different mechanical processes.

It is at the design and cutting stage that Marks & Spencer have started to employ young designers. The latter often have to extend their training so as to be aware of the mechanical processes involved in the mass production of fashion design.

Marks & Spencer are the UK's largest exporter of clothing. Apart from eight stores in France, Belgium, and the Irish Republic, the company also has controlling interest in many stores in Canada. Export customers sell St Michael goods in thirty other countries including Japan, Hong Kong, and Thailand, often setting up separate shops selling only St Michael goods.

The majority of Marks & Spencer's sales profits come from ladies wear, though these levels could now be at risk from outside competition. The ladies clothes department is under threat from a relatively new enterprise, Next, a company that has opened ladies clothing shops in many of the towns where there are existing Marks & Spencer stores. Next was founded in February 1982 from the original sixty branches of the Kendall rainwear business which was bought out by J. Hepworth & Son p.l.c. The original idea came from a great deal of market research, measuring the fashion market for up to twelve months. The evidence showed that there was a great gulf between Marks & Spencer – who aimed at an age range of 20 to 25 years plus and were renowned for quality merchandise but lacked design in general terms – and the top end of the sector, with Jaeger and Country Casuals, who gave quality and designer clothes but at high prices. The middle of this market was more or less untapped, so this was the area it was decided to attack.

Next's aims were:

(1) To give a range of merchandise suitable for the 20-plus market, the fashionable woman who wants to buy quality clothes.
(2) To inject a designer exclusiveness which is usually seen only in better-quality collections.
(3) To promote this at the right price ticket.

The Next range of merchandise is produced on a 'limited option' basis, so that limited styles are designed and manufactured to fit into a colour group, of which there may be up to six each season. Therefore, because there is a limited option, the company can afford to sell in volume. Thus, to the average customer the branch has an exclusive look with only a few garments in the particular selection; but with a very sophisticated replenishment system, heavy warehouse and stock-room stock, there is the back-up for volume sales. So unlike most retailers, Next sell from a very limited sales floor area and achieve higher productivity in terms of sales per square foot. The company is run on a complete designer collection basis. The design team choose the colour palette for the season ahead, and then design the clothes and choose materials. The design department is thus responsible for every garment in the collection. Next have very close links with several small manufacturers, which produce up to 90 per cent of their clothes. From the point of view of the way the company has grown, the manufacturers have been 100 per cent behind them, because Next have also secured their future. Over 80 per cent of their merchandise is bought in the UK, less than 20 per cent from abroad.

The establishment of Next has caused Marks & Spencer deep concern. In the past, Marks & Spencer felt there was no need to advertise their goods; everyone knew of the high Marks & Spencer quality. But because of the sales principles at Next, who co-ordinate their outfits down to the last accessory and display the outfits so clearly, Marks & Spencer have started to show greater concern in this area. Next publish a leaflet with each season's collection picturing their clothes and how to wear them. Marks & Spencer have now started publishing picture booklets too, as well as hanging large boards around the stores showing how their clothes can be co-ordinated. Next have proved that good shop layout and presentation do have a large influence on sales. If the outfit is presented with accessories, the busy working person will buy the package – saving time in shopping around. Unlike Next, Marks & Spencer have a less clearly demarcated store layout, but this is partly because of the size of the shops and the large amount of stock needed on the shop floor. Each department is kept separate – for example, knitwear, suits, skirts, blouses – but within departments positioning of stands does improve sales. For example, complementary gloves, hats, and scarves are placed together with winter coats in order to indicate to Marks & Spencer's customers possible ways of using accessories to enhance particular garments. The advent of Next has meant that Marks & Spencer have had to review their traditional practices in order to compete in the high street. But this has not necessarily meant rejecting some of the earlier company principles.

Despite the departmental competition there may be from outside rival companies, Marks & Spencer claim that over 14 million customers shop in their UK stores every week. The 'flagship' store at London's Marble Arch features annually in the *Guinness Book of Records* for taking more money per square foot than any other retailer in the world. This commercial success results from good quality, competitive pricing, and attention to detail in display, staffing, and design. This combination of ingredients has been revised and updated by the new generation of high-street merchandising chains such as Next, Habitat, and Principles, although they probably would not deign to admit it.

10

Presentation through display: shop-window dressing

PENNY RIDOUT

A shop window is a form of 24-hour advertising which expresses the philosophy of a shop and creates an identity. In areas like the King's Road in London, where there are many fashion shops, identity is all important. This is also particularly important for department and chain stores such as Debenhams, Marks & Spencer, Miss Selfridge, and Wallis. The window display establishes corporate identity so that whether a potential customer is in Leeds or London they can recognize the shop from its outward appearance. The ultimate aim of a shop window, then, is hard sell, showing merchandise which looks so attractive that people cannot resist going inside to find out more.

Recently the larger stores have started to display clothes in a more inventive way. This is partly because greater competition exists between the department stores, thus creating a greater need among individual stores to attract the public's attention. It may also be due to the fact that 'designer' shops, such as Katherine Hamnett's Norman-Foster-designed Garage in the Brompton Road, have drawn their attention to the previously untapped potential of shop display. As one would expect, each store has its own policy as far as window displays are concerned. However, the direction of displays as the responsibility of the staff varies from store to store.

The world of display is a creative world, as Adel Rootstein Display Mannequins Ltd describe it in their advertising pamphlet: 'A dream factory – where display designers, through their talents, produce dramatic, glamorous, three-dimensional atmospheres to induce, persuade and create a desire to "have" in the casual window shopper.' The right balance of atmosphere will hopefully reach out on to the high street and signal the indifferent passers-by to stop, look, and then go into the shop, where the sales staff take over.

Shop-window displays are nowadays becoming more and more abstract. Like fashion photographs and illustrations, they are becoming increasingly concerned with creating 'mood' or 'feeling'. For many shops it is now a case of the simpler the better. Simple, uncluttered windows with a strong theme are more striking and impressive than a hotch-potch of unrelated objects, which is important if you consider that a person's average attention span at any window is limited. In many department stores, mannequins are seen as an intrinsic part of the window display. Department stores spend vast amounts of money in continually updating their department's stock of mannequins and in renovating the make-up and wigs of their older models. On average, new ranges of mannequins are bought every year or so, and their make-up and hair-styles changed every four to six months. However, there are other ways of displaying clothes; for example, cardboard cut-outs and collapsed figures are sometimes used instead of mannequins.

Only a minority of shops, however, use these vehicles to display their fashions, the largest and most obvious being Next, who display some of their clothes on hangers, even in their shop windows. They often offset their displays with large fashion photos of the outfits on models, which together serve the same purpose as seeing the clothes displayed on mannequins. This then makes for comparatively inexpensive displays which lend themselves to being changed frequently and easily. Similarly, in many 'designer' shops such as Katherine Hamnett's Garage, and shops in the King's Road like Review, clothes are simply displayed on clothes rails, while the creative emphasis is put into the design and decoration of the surrounding building.

Window displays have been a static commodity for many years, and perhaps the future will see a kinetic form of window display more akin to live fashion shows. We are already seeing the installation of videos in many fashion departments, such as Harvey Nichols, showing companies' latest collections. As yet video screens have not been incorporated into shop-window displays, but maybe this will be the future of the art of window-dressing.

11
Wordly goods

SARAH TROTT

The fashionable image reproduced in the media often presents designs as 'new' when in fact they are revivals from the past or derived from older ethnic cultures. This was particularly evident in the 1970s. With the beginning of the 1970s, many younger people caught on to the burst of ethnic and hippie styles. The whole world was scanned for easy-to-make, easy-to-wear examples of peasant clothing. The idea was to use expensive basics, such as classic leather shoes, boots, Levi jeans, and *couture*-quality blazers or jackets, which were teamed with eccentric, 'way-out' items such as hand-embroidered Guatemalan or Mexican blouses, or with simple, effective, ready-to-wear items such as men's checked lumberjack shirts, large British Shetland jumpers, and Burberry raincoats.

There are many articles of clothing we have borrowed from other countries around the world which, when matched with tailored garments, for example, highlight the traditional construction of British fashion. Over the centuries we have helped ourselves to many useful and fresh ideas. Revived ethnic styles made a huge impact on fashion with the start of the 1970s. *Haute couture* espoused the peasant look and influenced high-street fashion; in turn it was acknowledged that street fashion had influenced the designers. The designers of fashion strived to parade something 'new' at fashion shows, but it was no surprise to find that the hippies internationally had influenced the catwalks of Paris and Milan. Ethnic dress was not merely a source of fresh inspiration to designers running out of ideas. It was an innovative idea which was picked up and utilized for the skill of the makers. The interest in embroidery or knitting from peasant sources is in part linked to the wish expressed by many people to escape from what they see as an artificial environment. This includes the preference for natural over synthetic fibres in clothes as a backlash against the synthetics of the 1960s. Nevertheless, until fairly recently natural fabrics have won only grudging acceptance in *haute couture* because they are more expensive to buy and, although more durable in the long run, are harder to take care of on a day-to-day basis.

The hippie movement adopted a more natural way of life. The hippies' ideal life was a reaction against the previous burst of all-out consumerism, the synthetics, and the general frivolity of the 1960s. Underlying this interest was perhaps an awareness that all things natural are not to be taken for granted. In general, the movement was towards a new simplicity, almost an austerity in style of dress, reflecting not only an awareness of the low quality of life experienced by people in poorer parts of the world and our exploitation of it, but also the economic difficulties facing the western developed countries. An interest in Latin America (for example) as a potential design source made way for the mass production of brightly coloured Peruvian-type weaves, and imitations were made readily available. More and more natural sources, like cotton, wool, and natural dyes, were found to cater for this growing taste, yet interestingly enough synthetics were sometimes used instead of traditional fabrics.

British fashion design was re-enacting a sort of nineteenth-century imperialism during the 1970s. In 1973 the United Kingdom joined the European Common Market, and also around this time jet travel brought the remotest countries into easy reach. There was increasing opportunity to travel further afield in what was becoming known as the 'global village'. These were reasons for fashion designers to be more aware of other countries. Anything from Eskimo apparel to gypsy costume was re-created, with such enthusiasm that collections seemed to have walked straight out of the pages of the *National Geographic* magazine. Perhaps fashion was being manipulated by the tourist industry. In addition, after 1962 it was noted that Russia was making a renewed effort in the field of fashion, and a recurrent ethnic theme followed. Four years later the film *Dr Zhivago* was released. Western fashion established itself firmly on the steppes of nineteenth-century Russia, with the aid of Yves Saint

Laurent's peasant collection of 1976. This also led to Diana Vreeland's exhibition, 'The Glory of Russian Costume', at the Metropolitan Museum of Art in New York. Ethnic revivals are often inspired and triggered off by major exhibitions.

A fashion revival is a regurgitating of ideas and trying them out in another context. It seems as though we are forever going back to the future, but this is partly because the designer has always got to know how the created object works and how previous examples were put together. From this basis a different interpretation of the original can be achieved.

Fashion is often considered derivative and seen merely as a vehicle for absorbing ideas from other countries, but what the fashion industry attempts to do is convert these ideas into a western idiom. In doing so the designs often become watered-down versions of the original. Symbols and particular patterns lose the original strength of meaning, but through being 'appreciated' by the west they take on an entirely different meaning. It is legitimate to a certain extent to take apart specific artefacts in order to learn different techniques of creating different fabrics, but this means that the articles will necessarily be more exclusive, since western mass-production techniques are not geared to the silk-weaving methods of the Japanese islands, for example. The versions of traditional dress that have been adopted by western fashionable society often display only whimsical interests or even ostentation. It is the job of the fashion designer to create a contextualized interpretation of the original. In the 1980s there seems to be an increase in the use of ethnic fabrics and costumes. As designers are we marketing a global culture in our use of such sources or exploiting references which are irrelevant to our own culture?

SECTION IV
Class and Clothes

INTRODUCTION

There are two approaches in this section. One considers the consumers of clothes in terms of their class background and their design preferences, and the other examines the social conditions into which garments are introduced and subsequently ritualized.

Many criticisms are levelled at the fashion industry, not least that many fashionable garments are expensive and thus accessible only to the rich. One essay argues that, because of the conservatism of many wealthy groups, they tend to experiment less with clothes than other social and economic classes and instead express their wealth through the acquisition of expensive, classic, and 'safe' garments. Money does not mean 'good taste' in terms of innovative design. One could argue that in one sense the restricted codes in terms of tradition and conformity inherent in the ruling class are more confining in terms of experimentation with designer clothes than are the restrictions due to the lack of money experienced by other classes.

There are types of activity which are engendered by the wealthy which consciously maintain tradition both in dress and in social discourse. One such activity was the 'afternoon teas' held in upper-class women's houses in the 1920s. The tea-gown itself originated in the 1870s and was at first a sort of dressing-gown which was worn at informal tea-parties attended only by women. By the 1920s the tea-gown had become a fashion garment in its own right. Fashion change was thus very slowly initiated by a ruling-class social function rather than by designers.

The first essay in the section (Chapter 12) looks at another ruling-class ritual – that of the débutantes of the 1950s and 1960s who were reluctant to transcend the prescribed rules regarding dress. The second essay (Chapter 13) considers their ancestors in the First World War and their children – the Sloane Rangers – who preferred and prefer to play safe with more traditional

Figure 7 Cover of *The Sunday Times Magazine Fashion Special*, 28 September 1986 (Photo: Neil Kirk/*Sunday Times Magazine*, London, © Times Newspapers Ltd, 1986). This cover does not project an image of a highly sophisticated, technology-based industry. The nostalgic overtones are harking back to when high fashion was dominated by *couture*, an elitist activity enjoyed only by a small percentage of society.

designs. The class-conscious nature of these essays illustrates the importance of a stratified social system in determining 'markets'. Investigation into popular and high-fashion codes and how they work is often a recognition of the relevance of class in a contemporary society.

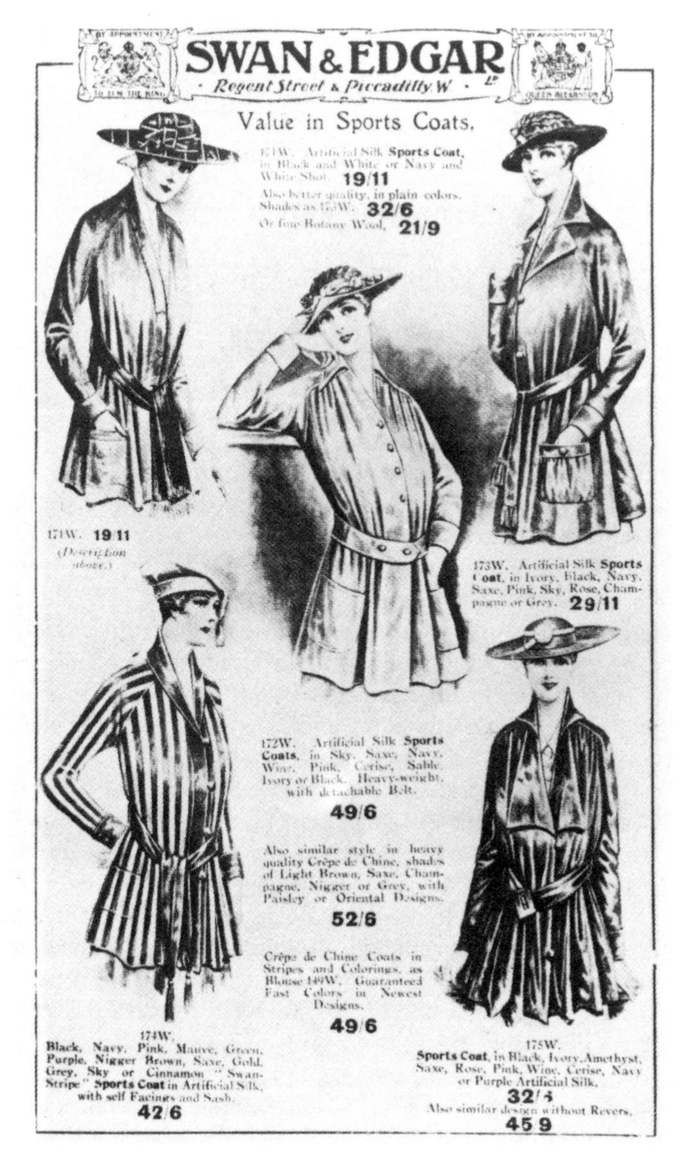

Figure 8 Sports coats (Swan & Edgar). Fashion was affected by the patriotic values of the First World War. Masculine forms of clothing were the precedent for women, who were taking over male occupations. Function began to take priority over fashion and tailored garments outdated the pre-war values of female fashion as conspicuous consumption. The long-standing tradition that fashion ideals start at the top and 'trickle down' was being eroded.

12

Debs or plebs: the disadvantages of being a privileged member of society

JANE SWANNELL

For most of us it is impossible to relate to the way of life of the ultra-wealthy and titled members of society, who, by their privileged upbringing and inherited wealth, have very little contact with the harsh realities of life. The frivolous life-style led by the débutante was a prime example of this. The ritual of young girls being presented at court for the first time, which we associate with twentieth-century British 'society', did not exist until the late nineteenth century. It was known as 'doing the season'. This practice entailed endless time-consuming devices and was taken on by a whole class.

The vintage years for debs were between 1930 and 1958, during which time the newspapers gave them star-like treatment. In 1958 the official presentations stopped, yet there were many rich parents who refused to accept that the débutante system had ended. Aided and abetted by the gossip columnists in the glossy magazines, they determinedly continued to describe their daughters as débutantes, having them curtsy to a large iced cake in Grosvenor House in lieu of the monarch, to mark the start of the season. 'Doing the season' is even now still too good an opportunity for the rich to be ostentatious for them to allow it to be dropped altogether. As the novelist Emma Tennant, a deb in 1956, says of the remaining breed doing the season, 'Everyone thinks the aristocracy is poor, but it is one big veil-pulling over the eyes. They are carrying on exactly as before. I don't see any difference between now, the 1950s and the eighteenth century' (quoted in *Vogue*).

For a girl making her début in society, a huge wardrobe of

clothes was needed for all the social functions at which she would meet the same set of people over and over again. Debs' mothers were snobs; they could not run the risk of their daughter being seen in the same outfit twice. Every precaution was taken to safeguard this from happening, and the deb was given an extensive wardrobe. By far the most important event of the season for the deb was the court presentation. From early history, it was the custom for subjects of the British Crown to wear their best clothes at important social and official functions, especially those associated with the court. Therefore, the court became the natural show-place for the most costly and elegant of clothes. Only the most reputable designers were considered worthy enough to supply fashions to the upper classes.

The media devoted a lot of coverage to the clothes worn by debs, who were often photographed in fashion magazines modelling expensive *couture* creations. Vast amounts of money were spent; wardrobes had to bulge with outfits worn fewer than a dozen times; and the press wrote ecstatically about these styles. Sometimes suggestions were made in special features about what should be worn for the season's events. Above all, conformity dictated the style of dress, whether it was for evening wear, day wear, for the country, or even sports wear. For the most part, fashion stayed within the restricted design ideas of the deb 'look'. The deb's appearance had become an institution and was restricted by orthodox views of how a young girl should look, especially when making her presentation to the sovereign and her début in society. Most debs were wealthy but did not patronize one designer in order to create an exclusive and original look. They used several *couture* houses, and the resulting image was always dictated by custom.

Debs were often opposed to the ladylike appearance imposed upon them, because it was a style associated with the older generation. But the dilemma could be solved for them only with the passing of years, because the restricted moral codes surrounding the clothes of the débutante were binding. While most of them wanted to keep up with the fashions of the day, the restricted design codes of the deb did not allow for flexibility. A deb's dress implied also a certain kind of demure behaviour. The need for formality often meant that debs dressed in a much older way than their contemporaries. They were not allowed to wear very young, iconoclastic clothes, neither could they be 'fashion freaks' without running the risk of alienating the entrenched 'safety' of the system.

Flicking through 'society' pages of *Harpers & Queen* or *The Tatler* in the 1980s is a sure reminder that, despite the waning of the official deb system, the upper class's dress sense, as far as evening wear is concerned, has hardly changed in thirty years. The strict rules for débutantes may have disappeared, the Honourables may be able to risk an Emmanuelle-designed evening dress, but clothes designed by the more avant-garde designers such as Richmond/Carnejo, Body Map, or English Eccentrics will rarely, if ever, be seen.

13

A class war: patriotism and fashion during the First World War

TIWAA ADJEPONG

Military wear influenced fashion greatly during the First World War. In 1915 many designers felt it necessary to transfer military ideas into fashionable garments. The 'officers' blouse', for example, became a craze for women and was made in a range of printed and plain fabrics. It was easy and comfortable to wear, with its own self-coloured braid on the cuffs and duplicate metal buttons. The fashionable version of the blouse was more fitted than the original official uniform, which had had regulation breast pockets and shoulder tabs. From this men's official shirt came many other fashionable blouses which were made by adapting the appearance but leaving the shape basically the same. Suggestions of adaptation were even published in *Woman's Own*, which had since 1914 become a favourite women's magazine. The 'military blouse' had gone through such a transformation that at the end of the war it had almost lost its previous identity.

Before the First World War tailored garments had been very much in demand, but they were less evident in the 1920s. For the more aristocratic classes, tailor-made clothing was very common, with fashionable coats resembling the man's military coat of 1914 (three-quarters in length); women's skirts, for practical reasons, rose above the ankle. The wealthier women of 1914 onwards were still able to afford to dress in the expensive country fashions of peaty tweed coat and clean-cut skirts. This type of outfit was often accompanied by a tailor-made shirt in silk or wool, completed with a separate linen or soft high collar and tie. Apart from the war influencing fashion in Britain, other countries, such as Russia, had a role in the change of styles and shape. One of the fashions in the first winter of the war came from the Russian trench coat, which was transformed in Britain into various colours of tweeds and gabardine, with, as accessories, thick waist belts made from the same fabric. The British women's single-breasted jacket was usually derivative of the men's military jacket, but was cut in a fitted form without back vents.

Whether influenced by the war or not, hats became a very important accessory of dress from 1914 onwards. The military caps and helmets of the war greatly influenced the appearance of fashionable millinery wear. The small close-fitting velvet toque, with its towering height and pointed wings produced in velvets, may well deserve its title as the 'skyscraper hat' and would be one of the best remembered from the First World War. Another popular type during the war was the flat sailor hats in silks, faille, or velvet trimings. Meanwhile, the Parisians were doing their bit to contribute to the new shapes in millinery as well as fashion garments by the use of decorative bows and ribbons daringly displayed. During the early years of the war, high Cossack caps in fur appeared to be in favour for an older generation of women. Small tippets of fur were worn to match and enhance the outfit.

Amongst the working class, where utility became the main criterion, a type of uniform austerity prevailed, leaving little room for the fashion-conscious. It became obvious amongst women workers in the factory that long hair as a fashion was no longer practical for the needs of war work. With the new cropping of hair a new style was set. There was a universal shortening of hair for both men and women, and by 1918 the *pompadour* hair-styles of the Edwardian period had completely disappeared. The boat-shaped hat and the *tricolore* were both designs influenced by the First World War which became high fashion amongst the spring collections of 1916.

When war broke out in 1914, it became clear that fashion in the *haute couture* sense of the word could not be sustained through the duration of the war. Munitions production took precedence over textile production. But there were those who, for whatever reason, would not give up lightly their dedication to fashion. While patriotic women believed in putting their country first, there were others who, despite the desperate seriousness of war, looked on fashion as a new and challenging aspect of life.

The summer of 1915 brought with it a crisis in the fashion world. Women endlessly discussed the importance of dress.

There were those war workers who declared that to follow fashion was to show a lack of patriotism; it was considered frivolous, in times of war, to put on new, full skirts and the latest fashions in frills and high collars. Others felt that it was the height of patriotism to adopt the new fashions and thus keep the mills and dressmakers in business. Many believed that the new fashions of the war would create new employment and enable scores of retail houses and dressmakers to carry on who would otherwise be compelled to close down.

One of the most noted uniforms of the First World War, which had a positive effect on the changes in women's fashions and attitudes, was the introduction of breeches. Trousers enabled women to have more physical freedom and be seen as more equal to their male colleagues. They meant that women could take up more active positions of work and feel less discriminated against, giving them the self-confidence to be able to pursue men's jobs during the war. Elizabeth Wilson, in her book *Adorned in Dreams*, describes the wearing of trousers by women during the First World War as 'Dressing for an expression of freedom'.

Trousers were not to be a completely acceptable mode of apparel for the majority of women until after the Second World War, whereas the dress and the skirt had been an established part of women's wear for centuries. Major changes occurred in women's clothing during the First World War, and yet after the war women's wear was to revert back to more traditional garments. The 1920s saw the loose 'flapper' dress but also the Paul-Poiret-inspired crippling hobble dress. There is a direct parallel between this fashion development and the way women who had worked in traditionally male jobs during the war were forced, when peacetime resumed, to return to domestic service and other traditional female jobs.

SECTION V
Specifics of Gender

INTRODUCTION

Clothes are usually designed to appeal to one gender rather than both by means of a conscious attempt to 'build in' those characteristics which are seen to represent that gender. The division is very marked in fashion design, where only a small percentage of garments are non-gendered. There are traditions in the genre of garments which are constantly reworked so as to be directed at the appropriate gender. The dress/skirt and the suit were established as suitable modes of dress for segregating the sexes at the end of the seventeenth century, and little has changed since. However, periodically this format has been challenged, as in the 1984–5 'men in skirts' theme or in the trouser-suit for women in the 1960s. Yet such innovation has never managed to overturn the predominant fashion forms.

This raises many questions as to the nature of clothing and its connections with socially motivated culture where gender differentiation is more important than integration. Visually dividing people by their physical characteristics can be traced back to classical traditions, when rules as to methods of clothing the body were established. One of the essays in this section (Chapter 14) looks specifically at attitudes to the *unclothed* body in Ancient Greece in order to attempt to establish how far societal moves affect our viewing of the human form and subsequently our dressing of it.

One of the basic functions of garments, whatever their type or style, is to cover the naked body, yet in the twentieth century a far more complex system is in evidence. Since the nineteenth century, changes of style have affected women's clothing more radically than menswear. The history of menswear is different to that of its female counterpart, since stylistic changes have not been as extreme as far as men's clothes are concerned. This warrants further investigation, since it indicates a type of gender

Figure 9 'Men as sex objects', a fashion photo appearing in *The Face*, September 1984. *The Face* magazine declared men as 'the new fetish, the new sex object' via a series of fashion photographs in 1984. One can speculate if this heralded a new era of gender equality or simply transferred exploitation to the other sex.

Figure 10 High street 'Casuals' (Photo: Marek Walisiewicz). Forms of rebellious clothing became Establishment after 1979, when anyone could purchase the punk uniform. The act of dressing up in the regalia with no life-style commitment undermined the rejectionist credibility. The new form was the visual appearance of convention which was subverted by action. The Casuals responded to an ideology which had outgrown itself and re-addressed the issue using vicarious means. In this illustration the Casual sartorial code of expensive European designer labels such as Lacoste has been translated into street fashion.

discrimination; that is, women are or have been so much more clothes-conscious than men.

Theories of fashion concerning female sartorial modes are not necessarily transferrable to men's clothing. Since there has been an overemphasis in historical and theoretical study on female dress, there is a need to redress the balance. The fashion industry is geared up to service a female consumer to a large extent. Perhaps this is because the style range of women's clothes has traditionally been broader, creating a larger market. Women have always been open to image manipulation and have raided the male wardrobe to extend its fashion potential. This is not an interchangeable process, and men have rarely utilized women's clothes to create a new fashion. If men do wear women's clothes this has often been read as effeminacy (see Chapter 15).

However, the notion that fashion is only for a female consumer is now being challenged. The menswear industry is becoming more aware of the potential of stylistic changes in men's clothes. In the post-war period, youth styles dominated the market, and subcultural fashions for men challenged the norm. It was pressure from these trends which succeeded in inspiring the industry to broaden the typology of male clothing (see Chapter 16). In the 1980s expanding menswear departments have emphasized casual clothing, which may have an indirect relationship to the subculture – the Casuals (Chapter 17) – who have come to be seen as a significant market force in the purchase of high-fashion leisure wear.

14
The unclothed figure

KEVIN CARRIGAN

Clothes are not only about dressing but also about undressing the human form whether male or female. Different perceptions of beauty in different societies have had particular implications concerning the garments which have covered up the naked body. In order to understand these implications as a fashion designer it may be useful to look back at a society whose perceptions of beauty were different from those entertained by the west in the twentieth century – to Ancient Greek society.

Narcissus, the Greek youth, was considered beautiful – so beautiful that it could be said that he wanted to make love to himself: a homophiliac? This beauty was more trouble than it was worth. Narcissus enjoyed hunting with his male friends – 'all men together' – but he was continually chased by females due to his beauty. One such female nymph, called Echo, had a past; as a member of the entourage of Hera (Zeus' wife), she was expected to be loyal. One of Hera's frequent activities was keeping an eye on Zeus' liaisons with nymphs, other goddesses, and mortal women. Though Echo owed her loyalty to Hera, when she saw Zeus indulging in his favourite pastime she warned him that Hera was not far behind. Hera knew of Echo's warning, and as a punishment the nymph was condemned to speak only when spoken to. She could echo, but she could no longer initiate. While Narcissus was being chased by Echo, he stopped for a drink by a pool and saw his reflection. He was immediately fascinated and there he remained transfixed. He leaned forward to kiss his reflection, but the face eluded his touch. He remained there, unsuccessful, and died. Taking pity on him, the gods changed him into a flower as beautiful as himself, casting its reflection into the pool. The flower still bears the name Narcissus as if to remind us of this tragedy.

Echo, who sinned through her voice, became just a voice. Narcissus became a beautiful but *passive* object – a useless but beautiful flower. Like a Greek sculpture, it waits to be appreciated. Its function is in the eye of the beholder. The stories of Narcissus and Echo are intertwined and they are also complementary. Echo suffered because she was a woman who took the active role; that was supposedly the man's role, and Narcissus did not behave as a man. He showed only one side of himself. As a punishment, the gods made him show his other side with equal vengeance. He was turned into what he despised – a woman, or classically 'passive'. We could contrast Narcissus to the 'he-man' stereotype, the total masculine male, as opposed to the totally effeminate male; an athletic, bronzed physique, always on the look-out for a female to boost his ego. To befriend females, however, to have them as close friends, supposedly emasculated men. Certainly many men lived as Narcissus, as we know from the position of women in Greek society, but a culture whose males prefer only other males is doomed to extinction. The Greeks' myth of Narcissus acts as an antidote.

The narcissistic male is instantly recognizable. Our explicit myth is: to be merely friends with women is to become one – effeminate, useless (in the eyes of 'real' men), waiting eagerly to be appreciated as an *object*. But at the same time, 'manliness' in excess is sure death, as the Greeks realized.

We have our own cult of 'real' men, just as the Greeks did: the one-sided 'he-man', preoccupied and narcissistic. Female beauty, such men think, is like the flower narcissus, passive. A 'real' woman waits to be asked, does not gallivant around with other men. Men are the active ones, having adventures, while the women wait at home. Men think they are more attractive if active – for example, the hero-worship of sportsmen. Narcissus was too effeminate because he was pursued and not the pursuer. Maybe his avoidance of women was because he was too much like them. Many men feel they have to prove themselves to other men, to be 'in' with other men, by not talking to women, instead making jokes at their expense. This could be because they may have difficulties getting on with women. A man compensates for this feeling by being 'manly' and doing 'manly' things. Getting back to the Greeks and Narcissus, the gods put a stop to this and made Narcissus *the object*, the supreme narcissist.

Beauty is a problem for women and for men. Men are presumed homosexual if they revere men. In women, beauty can cause jealousy and depression, and in some cases women who are thought of as beautiful want men to appreciate them for their intelligence as well as looks. For some men, beauty is still a great problem. Can they be comfortable with their own looks? Can they be comfortable with 'handsomeness' and beauty? If beauty in women is meant to be from head to toe, is it the same for men, or is their beauty located elsewhere? What is a good-looking man, to a man? Some can definitely tell you what they want in a woman. The Greeks had a clear view of what they wanted. Just compare the unclad, youthful, athletic male statues to those of the fully swathed female. Today we speak differently about men's and women's looks. Speaking about women we are able to speak of feminine beauty in many ways, each possible aspect covered in loving detail. We can speak about women piece by piece, organ by organ, but is it the same with men? Most women have very few words for their genitals, although men seem to have created quite a number for them; but talking with men about men the male organ is the centre of a description.

Like women, gay men frequently regard their own bodies as passive objects for the enticement and attraction of others. Hence, in a purely objective way, physical attraction is especially important – again, probably as important as to women. Gays have made a great deal of male beauty. For most gay men, there is no feeling of shame in the appreciation of other men and in vanity. They feel no shame about looking in the mirror and trying to look good. As being openly gay becomes more acceptable to the mainstream, gay and straight men will feel less pressure towards polarization, less compelled to accentuate their differences. But straight men, unlike gays and women, cannot compete sexually on the basis of looks alone. The heterosexual male is judged and judges by what he does. The image of man as a 'doer', man as worker rather than of leisure, has its roots in the past, the Greeks. Just as Aeneas was ready to sacrifice everything for destiny, or, like Odysseus, measured his worth by the power of his brain and of his 'cutting blade', from that time to the present, men with toil-roughened hands who work hard are thought of as intrinsically attractive.

For instance, take some of the beer commercials on television. Beer is sold by the means of a group of 'macho' men, hard at work, with toil-roughened bodies, but when they relax they deserve a well-earned beer and a woman. The advert is selling the image of male virility. It is also directed at the male sexual performance – that if he works and plays hard and drinks this brand of beer, a man's sexual performance will improve. By comparison to the beer advert, Levi jeans had an advertising campaign using macho males to sell jeans. Although these males were a new form of macho, they looked effeminate and vain, not working. Advertisements are generally sold through sex, whether male or female. Women are protrayed using their fingers and hands to trace the outline of the object to be sold, to cradle and caress its surface, teasing it as if it were a penis. Words like 'feel', 'deep', 'thrust', 'peak', and 'with muscle' are used to try to catch the mood of eroticism. The irony of the emphasis many adverts have of associating male beauty with work lies in the fact that in the late 1980s there is not much work to be had, for men or women. The question rarely asked in relation to high unemployment is whether with more enforced leisure masculinity will lose this traditional identity and become more 'effeminate', or whether the statuesque male will carry on being portrayed as an image unconnected to reality.

The word 'nude', and 'male nude', is really a taboo subject. Only in the art world and in art and design schools are we taught and educated to forget these inhibitions. Over the last 200 years, most artists have been obsessed by the female nude, yet for nearly 2,000 years the male nude overshadowed the female. It was the early Greeks who really appreciated the male nude. The male body was all important, and that period saw the evolution of a new naturalism based on new ways of seeing and feeling. We are now entering another period, with more men appearing in photography and advertising; artists are creating images, images they have studied, loved and experienced. Is the male nude once again going to be taken as the norm and the ideal? Are we going to see male nude full-frontals?

The naked body can sum up everything we desire and everything we most fear. The body is the source of our deepest pleasures and traumas; our whole experience of the world is set by the way we experience our bodies, by forgotten but all-pervading infantile fantasies. In our ambivalent attitude towards the body we are, unfortunately, heirs to our predecessors' beliefs – the Christians' denial of the flesh, and nineteenth-century reticence and repression. For all the talk over the last decade or so about the 'permissive society', few of us are really easy with the male naked body or even with the female. For some people, nakedness signifies liberation, a joyful and un-neurotic sexuality; for others, it stands to degrade traditional moral standards. Both of these seemingly contradictory attitudes rest on a common assumption, that the exposed body is emotionally charged and potentially subversive. If the naked body remains an effective sales gimmick in fashion advertising, pornography, on-screen, this can be only because we are still self-conscious about nudity, feeling that it breaks some still potent taboo. Although fifth-century BC Greece had a more healthy acceptance of male nudity, the Ancient Greeks had their own taboos, opposite to ours. They too allowed only certain kinds of nudity: the narcissistic nude male athlete as opposed to the well-draped female. They too were anxious about ageing bodies of both sexes, just as we are.

It is possible that the woman's body, whether aged or not, will always arouse more intense sexual response in men and women. Still in some men's eyes, the woman's body is expected to take a narcissistic pleasure in fulfilling men's fantasies rather than her own. There is still a rigid division between the sex that looks and the sex that is looked at, as John Berger indicated in *Ways of Seeing* (1972). During the last decade or so, feminists have questioned sexuality and the social divisions between the sexes. Men have traditionally had a large vocabulary to describe their sexual feelings about women; it is only recently that women have begun to express their sexual feelings about men. Women are capable of treating men as men traditionally treated women, as 'sex objects' or fetishized objects. But for women merely to imitate men and in reverse treat men as 'sex objects' is a perpetuation of the source process of alienating the nude from its sensual and human context. Magazines like *Playgirl*, mostly run by men, do this, making the assumption that women *want* and *need* male pin-ups and sex. But this reversal makes it none the better.

It is plainly impossible to will into existence a whole new world where there is no suppressed sexual feeling. Radical change never happens overnight. Maybe a revamped male and female nude imagery will have to wait until society's moral and wider political framework changes. Fresh perceptions of beauty will have to wait. If 'the new man' is to be more than a mere trend, if a more flamboyantly dressed male, regardless of sexual proclivity, is to emerge to be desired by women as well as men, then it will necessitate changes in the way images of the male nude are depicted in other areas besides the fashion industry.

15

Public image versus private self: the tragedy of camp dressing

KEVIN ALMOND

Camp dressing is a design for living that encompasses not only dress but the whole spectrum of artefacts that make up a personality. Camp is an excessive and highly theatrical form of life-style that involves both homosexuals, who exaggerate their behaviour because of social displacement, and heterosexuals, who perceive life through a homosexual context. There appear to be two essential qualities in camp: a secret within the personality which is alternately concealed and exploited, and a bizarre way of viewing life that is strong enough to impose itself on society through performance or creation. Camp people commit themselves to the marginal with a commitment far greater than the marginal requires.

The tragedy involved in being camp is varied. Some individuals are not in themselves camp, but the display of excess and tragedy in their life-styles invites the patronage of camp people. Others consciously adopt camp life-styles, bravely ignoring the inevitability of isolation and bigotry that any minority invites. Non-camp people are occasionally frivolous but mostly serious; camp people are only occasionally not frivolous! In a society as morally rigid as western society this frivolity is probably the only way to display ambiguous sexuality. The camp person wants to be accepted as he or she is, yet finds this to be unacceptable to the majority of other people. The public appearance becomes a self-advertisement of a private self or a mask that hides emotional or physical insecurities. Camp dressing can apply to both men's and women's apparel and yet it is more often used in association with men and men's behaviour.

The history of camp goes back a long way. But one of the periods in Britain in which it was most evident as a movement was the early and mid-nineteenth century. There was a severe difference between male and female dress then, men being both sober and colourless, women merely decorative and caged by constricting corsets. Dandyism originated as a reaction to such conformity of dress. Headed by such flamboyant figures as Beau Brummel and George IV, it was scorned as being both frivolous and effeminate. The Victorian moralist Thomas Carlyle, a fervent anti-dandy, attempted to define the movement in *Sartor Resartus* (1834):

> A Dandy is a clothes wearing man, a man whose trade, office and existence consists in the wearing of clothes. Every faculty of his soul, spirit, purse and person is heroically consecrated to this one object, the wearing of clothes wisely and well; so that as others dress to live, he lives to dress.

As with any minority, this camp dandyism was subject to the self-righteous patronizing of the majority. The Aesthetic movement of the 1870s, which boasted such camp-followers as Oscar Wilde and Max Beerbohm, was a cult dedicated to the beautiful in nature and fine arts. It was also revolutionary in that it altered the shape of women's dress. This was instigated by the Aesthetic painters, the Pre-Raphaelites, who depicted their ideal women dressed in the modes of the fourteenth century. The women of the movement began to adopt this medieval dress as a reaction to the corsets and whalebone crinolines that not only restricted movement but exemplified the caged and repressed role of women in Victorian society. The dress was basically a gown cut to appear loose and classical, made of limp materials in dull green, peacock blue, deep red, or yellow. It led one admirer to declare: 'Her dress is the ideal of what a dress should be, its colour, the fall of its folds, are soundless words set to music by herself.' Another stated: 'All the women look wan, untidy, picturesque, like figures of the Pre-Raphaelite pictures with unkempt hair.' Aesthetic dress was camp because it turned to the extreme of reviving ancient styles as a remedy for a drab world. It was revolutionary in its difference to contemporary female clothing, which exemplified the repression of women. Its own tragedy was that it not unnaturally suffered bigotry. *Patience* magazine described it as the 'Greenery Yallery Grosvenor Gallery' costume.

Camp is defined in the *Collins English Dictionary* as: 'Tents of army, military quarters, travellers' resting place,' The macho imagery of tents of army and military headquarters is a direct antithesis to camp dressing, yet it is the travellers' resting place for one type of homosexual view and some women's views of how to dress, act, entertain, and think. The world of camp is a fringe world of heroism not conventionally called upon to be heroic. Camp is a highly theatrical act, an exaggerated portrayal of reality. If it wasn't for the commercial and practical aspects of the fashion industry, one could quite easily see it, as a whole, as embodying all the ingredients of a 'camp' industry.

16

The new male: myth or reality?

KEVIN GENTLE

The image of the contemporary 1980s male has changed over the last few years through fashion design, written text, and the visual media. The present images are diverse and widely spread through various media channels. Magazines and newspapers as varied as the *Observer, The Times*, and *Cosmopolitan* have taken the example of John Galliano and Paul Smith, both menswear designers who, in projecting their own refined image of the male, provide a particular image for personal interpretation. Magazines, particularly the new men's magazines such as *Arena*, have had a specific part to play in constructing a variety of fashionable male images. Through the visual media the male has featured more prominently as a consumer of clothes in his own right rather than just as an accessory for the advertising of women's clothes. However, the latter still happens in the more up-market magazine adverts, such as appear in *Tatler* and *Vogue*, where an evening suit is generally still displayed by a 'gentleman', styled and stereotyped as the strong and silent type.

The image of the male in the 1980s is thus a rather confusing one, yet no more so than the plurality of female images with which women have been confronted for years. The essence of the 'new male', according to the *New Statesman*, is one who is becoming more self-conscious of what it is to be a man, and one who sees through the farce of masculinity and all the entrappings that accompany it. Unfortunately, it appears that this consciousness and fashion, which are the two key components of the 'new male', are rarely (if at all) brought together in the articles and fashion spreads in the media.

In the early part of 1985, when a great number of articles about 'the new man' appeared, the text, with accompanying photos, tended to state that men cared more than before about their personal appearance. Most articles then either proceeded to show a very outlandish mode of masculine attire and models with extravagant make-up, which only a minority would follow, or presented traditional outfits such as the suit. Neither of these two extremes of dress substantiated the text. This kind of attitude prevailed mostly in the mainstream fashion magazines (*Look Now, 19*, etc.), and merely served to state change in menswear rather than to discern what it was about new menswear designs that was different.

Menswear designers such as Jean Paul Gaultier, Scott Crolla, and Dean Bright used traditionally 'feminine' fabrics for their menswear, such as lace and silk, and mixed them with leather. Some journalists immediately assumed from this effeminization of menswear – particularly the 'men in skirts' syndrome of 1984–5 – that all men were at the point of becoming less aggressive and less stereotyped. In fact it was merely an interesting fashion story and could just as well have indicated the emergence of another 'new woman', in as far as these eroticized male garments were training women to sexually objectify men. Male fashions which initially, during 1984–5, got so much publicity through the use of 'feminine' fabrics and colours, such as the brocade and silk shirts and 'dandy' looks, have gradually become toned down to a more wearable and less obviously image-conscious look. The fashion images for men subsequent to 1984–5 have retained an air of experimentation but have not yet tended to be so extreme.

Newspapers and magazines have similarly dropped the label 'the new man'. He has become an old man with a new sense of style. Journalists are at times more fickle than the fashion industry itself.

17

Assemblage and subculture: the Casuals and their clothing

DEBORAH LLOYD

The Casuals came into public view in 1985 due to their involvement in football hooliganism. They are not a revivalist movement, but they do have a history which can be traced through three major influences – the first and most important being football culture, which enabled a select band of members from each club to create their own particular way of dressing, identifiable only to other members. This developed into rivalry between the supporters of different clubs culminating not just in violence on the terrace but in a kind of 'style wars'. Secondly, 'Northern Soul' was inherent in the development of football fashions, especially in the northern football clubs of Manchester and Liverpool. It was the attention to detail and the wearing of certain brands of clothes that the Casuals came to identify with. The 'Soul Boys' trend (*circa* 1976), instigated by David Bowie, was the third major influence. Their life-style encompassed the correct clothes, cars, clubs, and hair-styles, all of which changed in rapid succession.

These values led the way for the Casual movement in the 1980s. The Casuals, according to their appearance, are judged to be respectable. This low-profile appearance (as compared with skinheads, for instance) has allowed them to become a powerful force on the football terraces. Their respectable dress acts as a subversive disguise to the policing policies at matches.

The Casuals' sartorial codes are expensive, and the items they choose are not traditionally thought of as being affordable to the working classes. They are an upwardly-mobile culture, buying what they feel to be 'status' goods in order to improve their standing within their own community. The Casuals' idea of status is not the same as that of a middle- or upper-class subculture, such as the Sloane Rangers. They assemble their particular uniform, choosing certain items, constructing them into a collage, presenting a total look which is more than just an outfit but a way of labelling themselves so as to be identifiable as part of a group. The Casuals do not just take items of clothing from shops and wear them as intended. They adapt them like many subcultures have done before, as was pointed out by Dick Hebdige in *Subculture: The Meaning of Style* (1979). An example is the Teds, who took an Edwardian-style jacket and changed its colours and fabrics, thus altering society's attitudes towards it. This is something the Casuals are beginning to do with the items they choose, gradually bringing themselves into the public eye through the misuse of traditional pieces of clothing, not by physically damaging the clothes (as the Punks did) but by having subversive reasons for wearing them.

The Casuals are hard-core members of the 'unofficial' football supporters' clubs. They tend to come from urban areas with a strong football tradition which brings together a large percentage of the male community one day a week to support their team and defend their local territory. They are football supporters who have taken to wearing a form of dress which has nothing to do with team colours and traditional supporters' gear. Instead, they have adopted the idea that 'fashion is the code language of status' (Vance Packard, *The Status Seekers*, 1962) and have taken to wearing expensive sportswear labels such as Tachinni or Lacoste. Their uniform, however, is not static in its appearance; the whole basis of the Casual movement is change. The hard-core Casuals instigated this strategy in order to remain select, creating an élitist group which other supporters and clubs would try to emulate. The original Casuals did this by choosing expensive designer sportswear and changing the fashion from one brand name to another, ensuring the dedication of members and the select nature of the group through the tenacity needed to keep up with the trends. The fashions adopted by this subcultural group vary from club to club and within different regions due to slightly different influences.

It is uncertain exactly when this movement started, but on the Liverpool terraces in 1978 a trend was beginning. The emphasis was on detail; firstly some of the supporters decided to wear drainpipe jeans rather than just straights. Then it had to be Lois drainpipes. By 1979 the fads gained momentum. A new label was adopted every month; Inega, Fiorucci, Lacoste, and FUs all had their moments. No sooner was a brand name or colour adopted, established, and identifiable to the masses, than it was immediately cast off, replaced, and ridiculed. At Scotland Road, where the Liverpool supporters are based, inspiration in 1979 became thin on the ground. As a result, the style leaders adopted rather ridiculous trends by wearing no underpants, walking with their hands behind their backs, and even using Staffordshire bull-terriers as a fashion accessory. Casuals in London took to a luxurious image, an incongruous mixture of Nike trainers, frayed Lois jeans, Lacoste shirts worn with cashmere scarves and sweaters, often with the Pringle label, and long Burberry raincoats.

In 1980 a divide arose between the 'uniforms' worn by the different clubs, signalling the beginning of an ever increasing rivalry of style which culminated later in violence on the terraces. The 'Scallies' became identifiable by their Adidas ski anoraks, the Londoners by a variety of leather jackets and Lacoste, while Manchester's 'Perries' took to wearing anything with the Adidas label. By 1981 most of the major football clubs had their own collection of 'match dudes', each trying to outdo the next city in terms of 'terrace cool'. It is interesting that the sportswear labels taken up by the Casuals are not the English ones worn by their football idols, but foreign makes that are more expensive and difficult to obtain. However, this was the case only until retailers such as Olympus flooded the market, causing the style leaders to look elsewhere for more exclusive and expensive labels in order to outwit the masses. More recently the 'Scousers' have been returning to the mainstream of sports elegance. Media stars such as the character Damien Grant in the Liverpool-based *Brookside* highlight the new media awareness of a particular style, that style being Casual. In 1983–4 the standard 'look' for a London Casual was a 'wedge' or 'wet-look' hairstyle, often with highlights, and Fila or Ellesse tennis shirts; Gabicci and Cardin were *passé*, and Lacoste was on the way out.

As the hard-core trend-setters move on, only the dedicated can follow, needing money or tenacity (to steal the items) to keep abreast. It is not the retailers who ultimately dictate what the Casuals will wear, but they are responsible for the availability of garments, and companies such as Olympus are very quick to cash in on the trend by increasing supply in shops. An example of this is a small shop called Whitehall Clothiers, which seems to be typical. It is situated on Camberwell Road and originally was a school uniform shop selling blazers, badges, ties, and a little sportswear. In 1983 the uniforms were relegated to the back of the shop, later to disappear completely. The rest of the shop was then filled with expensive brand-name sportswear and casual gear: Lacoste, Fila, Diadora, Pringle, and Fiorucci. As a result, business started booming. The reason for this major change within two years was that the manager realized that certain lines of sportswear were outselling everything else. As a result they stocked up, and since then business hasn't looked back.

In 1985 another turning-point was reached concerning the London Casuals. They moved on from sportswear to designer wear – dropping the label on the outside for a more exclusive one inside the garment. These Casuals now patronize the top menswear shops in the West End of London, such as Browns, Georgio Armani, Ebony, and Woodhouse. This has caused problems for these high-class retailers, such as theft and the cutting off of the designer labels in the shop. The Casuals' clothes are not a visible insult in the way the Punks 'sex' T-shirts were, but they are an insult to the people they were originally intended for. This is especially true of designer wear such as Armani, originally intended for wealthy, style-conscious individualists who don't appreciate standing next to an East End lad wearing the same sweater. But the very fact that a Casual is wearing a designer sweater makes him look like he has the money and life-style that go with it – his 'fantasy self' – when in reality the sweater is probably stolen, bought on credit, or has been saved for over many months. They are trying to achieve status as opposed to class, because status can be bought, whereas class is socioculturally determined.

The uniform that the Casual has built up is a signal at football matches to both the home side and the opposing one, but has not become identifiable to the authorities such as the police. The Casuals have cleverly used their style to infiltrate football matches and continue to cause violence on the terraces. They have been responsible for a very particular type of football violence which has appeared in recent seasons. There has been an emergence of various inter-city 'fighting crews' connected with football clubs. These gangs have recognized names to cover their actions when clashing with rival supporters. Most of the names refer to and originate from the supporters who travel, not as part of the official supporters' clubs, but in advance on separate trains in 'style'. The Casuals' emphasis on style has helped outwit attempts to control football hooliganism. When they go to away matches, most supporters are met and herded by the local police force into the 'away' cattle pens. But a minority, the Casuals, will have travelled separately with little chance of detection. The recent ban on alcohol at matches will not have a great effect on this side of soccer violence, because drink at matches is frowned upon by Casuals; they feel it will affect their performance, that is, their fighting skills.

The trend began as a series of fads and fashions and escalated into a way of life. For some, following a football club has turned into a life-style – wearing the clothes, fighting to defend their team, territory, and pride. Theirs is a status-orientated life-style which is more than just fashion and football; it also encompasses cars, holidays, and the way leisure time is spent.

The reasons that the Casual movement hasn't been pin-pointed and documented to a great extent as a subculture are threefold. Firstly, the whole nature of the movement is change. It is self-perpetuating in that a select band of supporters from each club have taken to dressing in a particular way, making popular certain items or makes of garments. As soon as more than the select few catch on to this, the initiators change to another 'fad'. The Casuals' look is, therefore, difficult to identify at any moment in time. Secondly, although to the trained eye the Casuals are engaged in 'style wars', to outsiders they appear to be an undifferentiated mass. Thirdly, the Casuals are secretive about their dress codes and meeting places, which enables them to remain select.

The Casuals have taken to wearing a style of dress which wouldn't look wrong on a man of 50, and in some ways particular items have been stolen from the leisure wardrobe of the middle aged. Wearing such clothes seems to give them a veneer of respectability, an idea which the Mods adopted in the 1960s. To many, a subculture can just be a particular style of dress which can be adopted and disregarded without any knowledge of the history or reasons for wearing a certain uniform. This is one reason why cult styles often infiltrate mainstream fashion. The

Casuals are slightly different in taking clothing items from mainstream fashion and the high street, assembling them according to their own rules.

In most literature concerning subcultures, girls have been absent. Perhaps this has been because social reaction is concerned with the more extreme manifestations of youth subcultures. The popular press and media tend to concentrate on the more sensational incidents, such as 'Teddyboy killings', which are purely a male domain. In the case of the Casuals, the press has focused on football violence, which is also predominantly male, as is the rest of football culture. The Casuals tend to have a very sexist view of women. Women's importance within the movement is purely as an accessory not unlike the designer labels the Casuals wear. The girls have a minor role in football culture, but like the males they are interested in clothes and have followed the trend for designer wear. Like their male counterparts, status clothes and goods are important to them, which explains why a Casual's girlfriend whose fingers are 'decked in gold, is an important accessory. The Casual girls dress in a way which is aimed at attracting the male of their peer group. The Casual girl's wardrobe tends to be summer orientated whatever the season, consisting of many light, bright, and revealing items. These girls are rarely seen at football matches but are found in abundance in the 'designer' disco pubs and clubs. They play a minor but necessary supporting role.

The Casuals' form of dress is not particularly different from mainstream fashion; but the way they choose each item from different makes and shops, putting them together to form a 'look', makes them distinctive. True to the changing nature of their movement, they have started to move on from the label craze, deliberately removing the embroidered labels from the outside of their garments, but still wearing these items, conspicuous in the fact that the labels are missing. This shows that the Casuals adopt and then adapt their clothes. Another example of their customization is the re-emergence of flared trousers. The Manchester Casuals were responsible for this, and it can be seen as the logical conclusion to the development of their trouser styles over the last three years. Firstly, narrow jeans were split at the bottom of each seam to allow them to fit easily on top of the trainer; track-suit bottoms were worn with the side zips undone for the same effect. Both gave the impression of flares at the bottom of the trousers. The next step was actually to wear flared trousers. As a result, shops in Manchester have been cashing in on this style.

The Casuals' lust for status and style seems to be part of a general 'good taste' revolution happening on the high street. Shops such as Next and Habitat are offering designer life-style packages which are very inviting to the Casuals. This allows them to present a front to the outside world, believing themselves to be moving on from the working class and its previously associated trappings to something they believe to be middle class. This seems to be part of a general national *embourgeoisement* trend in the 1980s; many people want to be on a ladder upward at a time when the economic structure of society is directing them downwards.

SECTION VI

Manufacturing Images

INTRODUCTION

To a large extent the dissemination of information which reaches the consumer about fashion and what is fashionable is achieved through visual images. Photographic images are a representation of the real object but are used in such a way as to manipulate reality. To the designer, how their garment appears in a fashion feature in a magazine can be crucial, although it is not in their control. A media spotlight can lead to an increase in sales and can build or break reputations.

Despite its importance and abundance, fashion photography is considered a poor relation to photo-journalism, merely worthy of a flick through the pages of *Vogue* in a dentist's waiting-room. On the other hand, coffee-table books such as Helmut Newton's *Sleepless Nights* have elevated fashion photography to the level of a titillating art form. At its most basic level, fashion photography is a form of advertising for retail stores, designers, and businesses. On another level it represents the meaning of clothes in different contexts – shapes and cuts of garments, movement, garments in relation to human proportions, narrative, emblematic, nostalgic, or erotic. Varying techniques of fashion photography often illustrate contradictory moods for similar garments. Thus the received impressions of fashion photography, whether on the part of the consumer or the designer, may differ radically according to the representation such as glossy colour, black-and-white, hard, or soft prints. The first two essays in this section (Chapters 18 and 19) look at and attempt to analyse the role of fashion photography, the first schematically and the second more specifically in relation to two photographers working in black and white.

Photographic representations have dominated the media for much of the twentieth century, whereas fashion illustration tends to be relegated to working drawings. The 1980s have seen an upsurge of interest in fashion illustration as another form of fashion representation, both in advertising and as more personal expressions of fashion ideas. Together, photography and illustration have started to be used as a collage effect, and this development is the main concern in the last piece (Chapter 20).

As a result of there being so little material available which provides a critique of the fashion media, the work for this section is based almost entirely on the writers' own perceptions, observations, and analyses.

Figure 11 A fashion photograph appearing in *The Face*, September 1985. The combination of photography and a drawn image was pushing the boundaries of convention in fashion photography in 1985. Was it creating a new form of representation or addressing a trend which was particular to that moment in time?

18
Commercial images of fashion

LISA GOLDSWORTHY

Originally the basic purpose of fashion photography was to illustrate and sell clothes – advertising at its simplest and most direct. But over the years this basic requirement has been transformed into a subtle and complex operation that involves art, talent, technique, psychology, and salesmanship. To achieve its goal it needs to seduce the viewer into a world of glamour and illusion and in seeking to attain this it has chronicled and energized fashion, providing us with not only a visual history of twentieth-century fashion but also a vivid and lucid history of the images twentieth-century women have aspired and related to.

What is it about fashion photography that is so absorbingly seductive?

Fashion photography has always and almost exclusively been about and directed at women, but the images have almost always been created and controlled by men, and as such are inevitably suspect from a feminist perspective. On the negative side, this implies that fashion photography has been insidiously and exclusively manipulative, a powerfully and potentially frightening kind of brainwashing, a conditioning of ephemeral ends and misplaced values. In this respect the part that women fashion photographers play is obviously vitally important. If they choose to compete with men on their own ground (as several perhaps surprisingly do) the result, as in other fields, is even more aggressively masculine. On the other hand, a truly feminine approach can surely add another dimension to the field, giving it new authenticity and validity.

Fashion photographers on the whole appear to have had fewer

constraints, either economic or by the very nature of the medium, than their counterparts in, say, advertising or pornography, although sometimes the line has been so finely drawn it is practically non-existent. The creation of a fashion photograph somehow precludes reality; it calls for the invention of its own little world. However, although a fashion photographer may have his or her greatest successes within this field, it is often their other work, be it portraiture, documentary, or otherwise, that enriches and gives greater depth to their fashion work (for example, Cecil Beaton).

A complaint often levelled at fashion photography is that it doesn't fulfil its function and clearly depict and illustrate the clothes. This is where fashion photography breaks with commercial advertising and creates more than a realistic image of the object or garment. A fashion photograph does not represent reality, despite the fact that photography is generally considered documentary, a representation of something – a moment – that existed. A fashion photograph is not a statement of fact; it is an ideal, not commonplace reality but a created illusion. Although we all accept it as such, the medium works as a potent selling device, because subjectively people are willing to believe in the possibility of the existence that it depicts.

Fashion photography has suffered a great deal of criticism in its time: some justified, much unjustified, and little that cannot be aimed at fashion itself. The double stigma of commercialism and materialism, which it has in common with straightforward advertising photography, makes it one of the very few types of photography whose values are questioned, motives suspected, and aims despised. Its production for purely commercial purposes, despite its 'artistic' pretensions, implies creative manipulation and a sacrifice of artistic integrity.

However, the fashion photographer has a different role to that of the advertising photographer. Although the aim is ostensibly the same – to sell the product, the clothes – he or she is not usually employed by the manufacturer of the garments but by a third party, usually a magazine. Therefore he or she has far more room to manoeuvre, the magazine's constraints being a great deal less restrictive than the individual manufacturer's. The photographer is consequently able to exercise individual artistic talents to a far greater extent. In this respect, he or she is limited only by the fashion editor or art director they are responsible to for each particular session. The quality of fashion photography over the years has thus been as much in the hands of far-sighted and experimentally minded editors and directors as in the hands of the individual photographer – or even more so. Indeed, the reputations of many magazines depend on the quality of the fashion photography published under the reign of individual fashion editors, Diana Vreeland for *Vogue* in the 1930s, 1940s, and 1950s being perhaps the best-known example.

Various fashions may be acclaimed or derided, but the demand for new and different clothing is continuous. The public is not quite as easily manipulated as fashion critics would have us believe, and surely fashion's one sustaining feature is that clothing is partially a necessity, and one (like food and drink) whose qualities and refinements are appreciated by the large majority of people. It is a sensual appreciation and possibly self-indulgent, but none the less necessary for all that. Fashion design unquestionably has its own artistic devices: colour, proportion, spatial relationships, balance, and texture of fabrics being not least of its aesthetic qualities. The same applies to its representation through photography. Fashion photography presents a seductive suggestion and requires the viewer to aspire to an ideal, even if that ideal represents an unashamed portrayal of luxurious clothes, extravagant living, and exceptional physical beauty far beyond the realization of most spectators.

The consciously persuasive element – the manipulated identification of the viewer with the image, true of all commercial photography – may be one of the least attractive elements of fashion photography; but all of these negative aspects ignore its aesthetic validity. The best individual fashion photographs serve as works of art in their own right, quite apart from their fashion content. Besides the fashion photograph's ability to reflect so much of its contemporary social and cultural climate, which gives it a historical value, the aesthetic considerations involved are often as great as or greater than those in any other type of photograph, or indeed in any other two-dimensional art form. Although these aesthetic considerations are combined with a predetermined function – to illustrate and sell clothes – this should not diminish fashion photography's validity but give it an extra dimension.

The fundamental purpose of fashion photography and its *raison d'être*, the portrayal of fashion, may be almost incidental as far as the quality of the photograph and therefore its worth in terms of being recorded for posterity are concerned. However, the fashion content of the photograph should never be overlooked or underestimated, because although in the future today's fashion photography may be judged by different criteria, today it depends entirely on the successful and seductive representation of tomorrow's fashion, and that is what will decide whether it is remembered in the future. The fashion photograph cannot rely on the straightforward depiction of fashion; photography, being a two-dimensional representation on paper, is limited in terms of presenting the tactile sensuousness of three-dimensional clothing. But it can offer us a picture of the way the cloth moves and the garments hang, as well as more obvious qualities like colour, proportion, and fit. Like fashion illustration or any other artistic interpretation, fashion photography has to use its own potential to add to the individual subject-matter in hand.

19

Black-and-white fashion photography as a medium of expression

JACKY MARSHALL

If feelings are to be expressed through photographs, the photographer has to know the many different techniques which can be used to create moods and impact. This essay looks at two photographers who are established in the field of fashion photography and who are instantly recognized by their use of individual techniques, especially in black-and-white photography. Their work is different, yet they have the same desire to create photographs that are graphic, emotional, and fascinating to look at.

I want to explore how black-and-white fashion photography can, in certain photographers' work, incorporate so many ideas relating to reality (clothes), mood (art), technique (photography) in a moment, and sentiment (emotional permanence).

Deborah Turbeville is a former fashion editor turned photographer. Her career began as fashion editor to *Ladies Home Journal, Harpers Bazaar*, and the short-lived magazine *Diplomat*. She was invited to a photographic seminar under the direction of Richard Avedon and Martin Israel, which led to a job as photograph editor on *Mademoiselle* in 1968. In 1972 she left the United States and went to Paris, where she now lives. She does fashion assignments for *Vogue Italia, Marie Claire,* and *Nova*, and has done advertising campaigns for Valentino and Charles Jourdan. Her photographs broke all the traditional rules of the 1970s (contrasting sharply with the creamy prints of fashion photographs then – the kind produced by large cameras and complicated studio lighting) and would not have been possible without the radical changes introduced by Guy Bourdin in the

1960s. He changed fashion photography from a static concept to that of a narrative one, and Turbeville's contribution was to extend Bourdin's ideas. She had an intensely personal style. She used odd colours and tone, fantastic images, ironic humour, and off-beat sex, working specifically with models and situations that had depth and character.

The famous 'Bath House' series of photographs for American *Vogue* (May 1975) signalled Turbeville's arrival as a major fashion photographer. The photographs were described as evoking the grisly aura of a concentration camp or the frightening vacuousness of drugged stupor. The figures are isolated, combining passive, laconic poses with expressions of unease and disenchantment. It is as though each model is frozen in space, without any eye contact with each other or the viewer. These photographs were really quite disturbing and they were one of the most controversial fashion spreads ever published. The slouching posture seen in these prints has become one of the marks of Turbeville's style. Intensified by the lack of eye contact between models and the camera, the slouching posture of the models reflects the changed moods and attitudes of today's young women. This was the turning-point in Deborah Turbeville's stylistic development.

She creates narrative work, containing surrealistic and cinematic qualities. The feeling and sensualilty that the 'actors' in her photographs show is also her own insight into the pain and vulnerability of modern woman. She herself has said that the photographs are like the women you see in them: a little out of balance with their surroundings, waiting anxiously for the right person to find them. The women in Turbeville's photographs play an important role, and when critics write about her work they always devote a lot of time describing the slouching poses, anorexic bodies, and delicate pathos that the models radiate.

The traditional tendency of fashion photography in the 1970s had been to reduce women to erotic objects. This tendency is missing in Turbeville's work. She likes women and does not fear them. She chooses to use their inner qualities rather than using models who smile glossily into the camera. Her women are elusive and lonely, with brooding, anxious expressions. Their deep-set eyes do not sparkle and contain no flirtatious come-on or put-down. It is the whole woman who dominates the scene. Her locations have the same aura as the models. You are always aware of the air around them. Outdoors, they peer at us through mist and fog; inside they wander through empty space in light filtered by dirty and cracked windows. These empty corridors replace the seamless backdrops and elegantly furnished society mansions that we are accustomed to. Her settings are more like a mystery play. The misty, grainy style she has adopted evokes another time, an imaginary world. This is contradictory to the stereotype fashion photograph of clear black-and-white images.

Turbeville combines tones and texture with the overtones of foreboding. She scratches and marks her prints, distorting the image even more, or photographs images through broken windows so that we can only glimpse at what is behind the opaque surface. She creates anonymous settings; the mood is of menace and uncertainty; muted tones enhance the feeling of remoteness. Her printing techniques are as well thought out as her choice of models and location, and contribute as an essential part of the photograph and atmosphere. The clothes are hardly the meaning of these pictures. She is more interested in making eye-stopping prints. Only occasionally does she give us a glimpse of detail, as with the shoes. Generally it is difficult to get a clear idea of the clothes, let alone any detail. Yet the magazine reader's eye is caught. The women are real and appealing, and the reader can identify with them.

Bruce Weber is the second current fashion photographer to be considered, in order to contextualize Turbeville's style. In general his work shows a preference for natural light. This is evident from many aspects of his work. His outdoor shots are more poignant than his studio work. They have more feeling and mood, due to his mastering and control of different types of daylight. This in itself is a great achievement, since the whole concept of photography is based on the use of light. Weber has a particular skill and knowledge, which enables him to be stimulated by the natural elements. This spark of sensitivity shows in his outdoor black-and-white photographs, which are also noticeably more relaxed and less contrived than his studio work. He also used sepia toner and other colour tints in his photography; without turning it into colour, sepia 'warms' all the tones to a browny colour, making the photograph look old and worn. Weber's prints are of an all-over even tonal quality which is very well suited to the sepia toning method. The sepia adds to the mood and atmosphere that he tries to capture, especially in his American pioneer prints.

The weather and location play an important part in setting the mood in Weber's work (as do models and his unorthodox treatment of clothes). In one photograph in the series titled 'Homage to Edward Weston' he captures the sea air and the dampness, giving the photograph mood and telling a story. It could be said that he is giving us the obvious and leaving nothing to the imagination; all his images are very clear, and it is partly because of this that his photographs lack any real depth. There is no taxing of the brain, no mystery (unlike in Deborah Turbeville's work), but despite this his photographs are very pleasing to the eye and unusual in their approach to fashion. His sensitivity to nature has made his outdoor work famous, and he readily admits that his influences are photographers with the same ability to capture the elusive quality of the elements.

'Homage to Edward Weston' is a series of photographs done For British *Vogue* (December 1982), as was 'In an English Country Garden', which was undertaken in the spirit of an admiring tribute to the work of Cecil Beaton. In both cases, the influence is indirect. Weber evokes an atmosphere rather than referring to specific images. In 'Homage to Edward Weston' he tries to use the same idea as Weston by using his relationship with nature to create photographs which give the impression of being uncontrived. These photographs stimulate one's senses and memory in a very harmonious way (especially when compared with the misery and despondency of Turbeville's work).

The range of Weber's influences are vast, but ultimately it is the originality with which he interprets his own ideas that makes him a key figure in fashion photography today. His fashion work is of a very high standard, and he has contributed to many of the top magazines, such as British *Vogue*, *Tatler, Vogue Bambini,* and American *Vogue*, as well as advertising for top designers like Karl Lagerfeld, Ralph Lauren, and Calvin Klein. An interesting series of photographs was commissioned for *Vogue Bambini*, the main children's-wear magazine in Italy and the world. Instead of photographing beautiful children in expensive dresses, he did a narrative called 'Camp Rowdy', producing a 'fun' set of photographs of kids having a cheerful time dressing up and playing at being camp. These pictures are action-packed with children, and Weber manages to create the mood of all those smelly and dirty children really enjoying themselves away from their parents. He makes use of real people in place of models, incorporating groups (as with the 'Camp Rowdy') and even nudes in otherwise conventional fashion shots.

As with Deborah Turbeville, Bruce Weber feels that he isn't specifically a 'fashion photographer', because many of his photos have a strong element of portraiture and narrative. He says that there is no distinction in his own mind between his fashion photographs and the kind he exhibits in galleries. Deborah Turbeville has said the same thing, although she seems to go a step further. Whatever she is, she combines portraiture, photojournalism, and fashion photography in her work. That is the beauty of fashion photography today. There aren't many set rules to follow, and you create what you want using photography as a medium like paint or any other material – although photography differs from paint in its invaluable ability to capture moments in time. Weber's influence extends beyond technique. He was one of the first exponents of the natural unmade-up face and helped transform an appreciation for early American clothing into the 'prairie look' and 'pioneer look' that became so popular. He prefers clothes which suggest something arbitrary, worn, mismatched, and unpredictable, grace born upon chaos. He also turned his admiration for the long hair and muscles of surfing and athleticism into a force which revolutionized male modelling.

Bruce Weber's photographs of men contain the same knowledge of the male that Deborah Turbeville's photographs have of the woman as a haunted, desperate, interesting, and beautiful character. It is in the portraits of athletes that Weber's true feelings emerge. The prints possess an intensely personal expression of his feelings towards men (as with Deborah Turbeville's Bath House series, which say something about the way she feels towards the women in the photographs). It isn't only Weber's photographs of men that have this extra quality, but also his fashion/portraiture work of a model called Talisa. She occurs in many of Weber's photographs through his choice. Like many photographers who have favourite models, he obviously enjoys photographing her a great deal.

Bruce Weber has observed people and his surroundings, training his eye to note everything he sees, even if it means observing other photographers' work and technique and using them to his advantage. His vision is intensely personal. Techniques are ultimately important, not only in producing a good quality print but also in producing a style that is recognizable as an individual photographer's, as in the case of both Turbeville and Weber.

20

Photographic and illustrative fashion representation

JULIAN SMITH

When fashion was first seen in magazines, illustration was the key medium. Over several decades, however, photography almost made illustration redundant. But the balance between fashion photography and illustration has been swinging back in favour of illustration in the last two or three years. Why has this occurred?

Images in fashion photography during the last fifty years have changed as a result of a variety of technological photographic advances. In fashion illustration, there has been a progression from realism to expressionism which seems to have been a major change in the twentieth century. Combined with this there has been a change in rendering and printing techniques. Currently, black-and-white photography seems to be making a more noticeable return. Early fashion illustrations were monochrome or very simple line because of printing limitations; the availability of colour reproduction and printing was to follow. The recent return to black-and-white photography indicates a possible step back towards line drawing, despite the fact that technology allows for complex images to be created.

In representing fashion visually, illustration and photography have many different qualities. Because of these varying qualities, they each focus on different aspects of fashion. Photography pays particular attention to the environment, to bodies (possibly nude), and to glamour, whereas illustration tends to cut out the surrounding information in its representation and emphasizes the garment.

Most fashion magazines in the 1980s contain at least 50 per cent of advertising material. Fashion advertising has developed from 'straight' photographs, taken by a photographer employed by the garment producers, to purely individually constructed photographs accompanied by footnotes about the clothes. With this development there seems to be a division in fashion photography between 'pure' fashion photographs and the more 'artistic' photograph doubling up as an advert. Fashion is about clothes; but fashion photography involves not only the clothes but also the body. This use of the body in some photographs seems to overtake and overpower the clothes. By focusing on the body, fashion photography can contradict fashion, because fashion itself is about clothes making a body desirable or changing its appearance.

Fashion illustration for advertising seems to be used mostly for expensive garments appearing in glossy magazines. However, the cruder style of illustration is developing in magazines like *The Face* and *ID*. Fashion advertising is constantly striving for new methods of approach to representation, which is why in the mid-1980s collages of illustration and photography started appearing in some of the more avant-garde fashion magazines. One of the problems with this sort of image is an organizational one of involving two artists who use two very different media in the production of just one image. This might prove difficult as far as temperaments are concerned, but also might mean that the art directorship of a particular advertisement would outweigh the managerial directorship. To most advertising managers, as long as the clothes are seductive in a way which creates a saleable image, attractive to the viewer, then the picture has achieved its function. For the illustrative artist and photographer the aesthetic appeal of the image becomes as important.

Fashion illustrators, such as Antonio Lopez and Tony Viramontes, have begun to be acclaimed in a similar way to top fashion photographers like Bruce Weber. Were they to converge their skills in an unholy alliance, photo-illustration could produce some of the most interesting fashion advertising to date. Instead of fashion illustration superseding photography, the way forward may lie with the bringing together of the two media – which could be called the 'third field'. The combination of technology (photography) with fine art (illustration) is bringing a new expressive form to the representation of fashion.

Suggested reading

SECTION I

Brandon, R. (1977) *Singer and the Sewing-Machine: A Capitalist Romance*, London: Barrie & Jenkins.
Burton, Anthony (1984) *The Rise and Fall of King Cotton*, London: BBC/Deutsch.
Chapkis, Wendy, and Enloe, Cynthia (1983) *Of Common Cloth: Women in the Global Textile Industry*, Amsterdam and Washington, DC: Transnational Institute.
Dickens, Charles (1977) *Hard Times*, London: Pan Books, reprint.
Engels, F. (1979) *The Condition of the Working Class in England*, St Albans: Panther, reprint.
Fishman, William (1975) *East End Jewish Radicals*, London: Duckworth.
Harvie, C. (1970) *Industrialization and Culture 1830–1914*, London: Macmillan.
Pinchbeck, Ivy, (1981) *Women Workers and the Industrial Revolution 1750–1850*, London: Virago.
Samuel, Raphael (1975) *Village Life and Labour*, London: Routledge & Kegan Paul.
Thompson, E. P. (1968) *The Making of the English Working Class*, Harmondsworth: Penguin.
Walkley, Christina (1981) *The Ghost in the Looking Glass: The Victorian Seamstress*, London: Peter Owen.

SECTION II

Booker, Christopher (1969) *The Neophiliacs*, London: Collins.
De Marly, Diana (1980) *The History of Haute Couture*, London: Batsford.
Glyn, Prudence (1982) *Skin to Skin*, Hemel Hempstead: Allen & Unwin.
Hollen, Norma and Saddler, June (1980) *Textiles*, London: Macmillan.
Kunzle, David (1982) *Fashion and Fetishism*, New York: Rowman.
O'Neill, Elizabeth (1979) *9½ weeks*, London: Sphere.
Parker, Roszika (1984) *Subversive Stitch*, London: Women's Press.
Pollock, Griselda, and Parker, Roszika (1981) *Old Mistresses: Women, Art and Ideology*, London: Routledge & Kegan Paul.
Stone, Elizabeth (1841) *The Art of Needlework*, London: 3rd edn, now out of print.
Willis, Paul (1978) *Profane Culture*, London: Routledge & Kegan Paul.

SECTION III

Adburgham, Alison (1981) *Shops and Shopping*, Hemel Hempstead: Allen & Unwin.
Carter, Angela (1982) *Nothing Sacred*, London: Virago.
Douglas, Mary, and Isherwood, Baron, (1979) *The World of Goods*, New York: Basic Books.
Forty, Adrian (1986) *Objects of Desire*, London: Thames & Hudson.
Frazer, W. Hamish (1981) *The Coming of the Mass Market*, London: Macmillan.
Herdeg, Walter (1982) *Window Display – An International Survey of the Art of Window Display*, London: Sylvan Press.
Jabenis, Elaine (1983) *The Fashion Directors*, Chichester: J Wiley, 2nd edn.
Midgley, David and Wills, Gordon (1973) *Fashion Marketing*, Hemel Hempstead: Allen & Unwin.
Miller, Daniel (1986) *Material Culture and Mass Consumption*, Oxford: Blackwell

SECTION IV

Baines, Barbara (1981) *Fashion Revivals*, London: Batsford.
Flugel, J. C. (1950) *Psychology of Clothes*, London: Hogarth Press, reprint.
Glyn, Prudence (1978) *In Fashion*, Hemel Hempstead: Allen & Unwin.
Hall, Radclyffe (1976) *The Well of Loneliness*, London: Barrie & Jenkins.
Konig, Rene (1973) *The Restless Image*, Hemel Hempstead: Allen & Unwin.
Rowbotham, Sheila (1972) *Women, Resistance and Revolution*, London: Allen Lane.
Sparke, Penny (1986) *An Introduction to Design and Culture in the 20th Century*, Hemel Hempstead: Allen & Unwin.
Veblen, Thorstein (1957) *Theory of a Leisure Class*, Hemel Hempstead: Allen & Unwin.
Wilson, Elizabeth (1985) *Adorned in Dreams*, London: Virago.

SECTION V

Anger, Kenneth (1986) *Hollywood Babylon*, London: Arrow.
Barthes, Roland (1985) *The Fashion System*, London: Cape.
Burnett, T. A. S. (1983) *The Rise and Fall of a Regency Dandy*, Oxford University Press.
Carlyle, Thomas (1984) *Sartor Resartos*, London: Dent Everyman, first published 1834.
Cohen, Stan (1980) *Folk Devils and Moral Panic*, New York: St Martin's Press.
Crisp, Quentin (1977) *The Naked Civil Servant*, London: Fontana.
Hebdige, Dick (1979) *Subculture: The Meaning of Style*, London: Methuen.
Packard, Vance, (1962) *The Hidden Persuaders*, Harmondsworth: Penguin.
Walters, Margaret (1979) *The Male Nude*, Harmondsworth: Penguin.
Williams, Raymond (1976) *Communications*, Harmondsworth: Penguin, 3rd edn.

SECTION VI

Barthes, Roland (1977) *Image, Music, Text*, London: Fontana.
Barthes, Roland (1982) *Camera Lucida*, London: Cape.
Hall Duncan, Nancy (1979) *The History of Fashion Photography*, New York: International Museum of Photography/Alpine Book Co.
Hayward Gallery (1986) *Surrealism in Photography: 'L'Amour Fou'* (catalogue), London: Hayward Gallery.
Newton, Helmut (1978) *Sleepless Nights*, London: Quartet.
Packer, Bill (1980) *The Art of Vogue Covers*, London: Octopus.
Sontag, Susan (1979) *On Photography*, Harmondsworth: Penguin.
Steele, Valerie (1985) *Fashion and Eroticism*, Oxford University Press.
Victoria and Albert Museum (1985) *Shots of Style* (catalogue), London: V. and A.
White, Cynthia (1970) *Women's Magazines 1963–68*, London: Michael Joseph.
Widgery, David (1986) *Beating Time*, London: Chatto & Windus.

Index

Note Page numbers in **bold** type refer to illustrations.